FREAKY
SCIENCE

FREAKY SCIENCE

1,500 weird and wonderful scientific facts

MARK FRARY

METRO BOOKS
NEW YORK

Designer: Dave Ball

See page 128 for picture credits

Metro Books
122 Fifth Avenue
New York, NY 10011

ISBN-13: 978-1-4351-0867-7

Conceived and produced by
Elwin Street Limited
144 Liverpool Road
London N1 1LA
www.elwinstreet.com

Printed and bound in China

10 9 8 7 6 5 4 3 2 1

Contents

THE VERY SMALL

SMALL

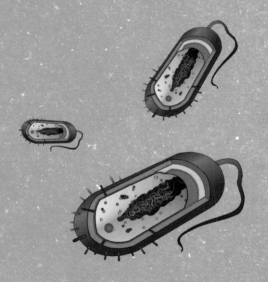

The language of numbers

Scientists love using special numbering systems and names for really small and really big numbers to bamboozle everybody else. The problem is that they come across the microscopic every day and when they talk about numbers or plug them into equations, it can be hard unless everyone is talking the same language.

Mathematical shorthand

Consider an atom of hydrogen. It measures about one ten millionth of a millimeter across. If you write it in meters it looks like: 0.0000000001 meters. Scientists writing sums like this would soon run out of paper in their notebooks, so instead they use a mathematical shorthand.

Prefix	Symbol	Fraction/multiple	Powers of ten
atto–	a	quintillionth	10^{-18}
femto–	f	quadrillionth	10^{-15}
pico–	p	trillionth	10^{-12}
nano–	n	billionth	10^{-9}
micro–	m	millionth	10^{-6}
milli–	m	thousandth	10^{-3}
centi–	c	hundredth	10^{-2}
deci–	d	tenth	10^{-1}
deka–	D	ten	10^{1}
hecto–	H	hundred	10^{2}
kilo–	K	thousand	10^{3}
mega–	M	million	10^{6}
giga–	G	billion	10^{9}
tera–	T	trillion	10^{12}
peta–	P	quadrillion	10^{15}
exa–	E	quintillion	10^{18}
zetta–	Z	sextillion	10^{21}

We know that $100 = 10 \times 10$. Instead of writing 100, scientists write 10^2 and say 10 to the power 2. Equally we know that $1,000 = 10 \times 10 \times 10$ or 10^3 or 10 to the power 3. A million is 10^6. A similar system is used for small numbers. A hundredth or $1/100$ is the same as $1/10 \times 10$ so scientists write this as 10^{-2}. Similarly, a millionth or $1/1,000,000$ is written as 10^{-6}

Scientists also use prefixes when talking about measurements. Some of these you may have already seen: the milli- in millimeter, the micro- in microsecond and the giga- in gigabyte, for example.

Radioactivity and how it was discovered

At the end of the 19th century, the German scientist Wilhelm Röntgen was doing some tests using a discharge tube – a glass tube containing electrodes which are used to pass electricity through different types of gas within the tube. By chance he noticed that photographic plates some

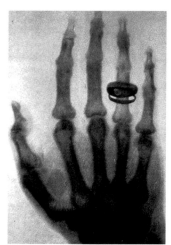

distance from the tube glowed when it was in operation, even if the tube was covered with thick, dark paper. He found that glow varied according to the thickness of the object placed between the tube and the photographic plate. When he put his wife's hand between the two, the image on the plate clearly showed her wedding ring and the bones in her flesh. Röntgen had discovered X-rays.

LEFT The hand of Wilhelm Röntgen's wife, displaying X-ray for the first time.

Radical radiation

X-ray was the first discovery of what is now commonly known as radioactivity – a process whereby atoms spontaneously emit radiation or a stream of particles. The other forms of radioactivity, all found by the beginning of the 20th century, are:

Alpha (a) particles: *Discovered by Ernest Rutherford from studying uranium. Originally thought to be radiation, they were found to be particles made up of two protons and two neutrons. Typically emitted from radioactive elements at low speed and easily absorbed by other materials. Strongly ionising – can easily strip electrons off other atoms. Naturally occurring uranium and thorium are strong alpha-emitting radioactive isotopes (different forms of a particular chemical element, distinguished by the number of neutrons contained in their nuclei).*

Beta (b) particles: *Also discovered by Rutherford, these were originally considered to be some kind of ray, but are in fact a stream of energetic electrons. Beta particles are highly penetrating and can be emitted from radioactive elements at close to the speed of light. Strontium 90 and carbon 14 are both strong beta emitting radioactive isotopes.*

Gamma (g) rays: *Discovered by Paul Villard in 1900 (but named by Rutherford). Villard recognized they were similar to Röntgen's X-rays but had a far greater penetrating power. We now know they are, in fact, another form of electromagnetic radiation, like X-rays, radio waves and visible light. The isotope cobalt 60 is often a source of gamma rays.*

Rutherford also discovered a property of radioactive isotopes – half life. This is the amount of time over which half of an amount of an element will be transformed into other things through the processes of radioactivity. It can range from just a fraction of a second to, in the extreme, billions of years.

Radioactive elements and their uses

Element	Half life	Uses
Americium 241, 241Am	432 years	Smoke detectors, thickness gauges
Polonium 210, 210Po	138 days	Satellite power source, believed to have been used to poison former Russian spy Alexander Litvinenko
Radium 226, 226Ra	1,600 years	Discovered by Marie Curie, used to calibrate detectors
Radon 222, 222Rn	3.8 days	Radiotherapy
Strontium 90, 90Sr	29 years	Radioactive power sources, radiotherapy, bone cancer treatment
Technetium 99m, 99mTc	6 hours	Tracer in medical diagnosis
Thorium 232, 232Th	14 billion years	Source of fuel in breeder reactors, radioactive dating
Uranium 235, 235U	700 million years	Nuclear energy generation, nuclear bombs
Uranium 238, 238U	4.5 billion years	Source of fuel in breeder reactors, radiation shielding, nuclear weapons
Ytterbium 169, 169Yb	32 days	Brain scanning, radiotherapy

What do atoms look like?

The Greek view

For two thousand years since they were first conceived by the Ancient Greeks, atoms (from the Greek atomos, meaning indivisible) were considered to be tiny featureless balls that were the fundamental building blocks of all nature. Each ball was thought to be the same but combined in different ways to create the different things they saw around them.

Dalton's view

Not everyone agreed. A new atomic theory put forward by English chemist John Dalton in the early 1800s achieved acceptance more quickly. He said that ultimately each element was composed of atoms but that atoms of one element were different from those of other elements. In particular, they had different weights. As a result, the alchemist's dream of turning other elements into gold seemed doomed to failure.

The negative discovery

Less than a hundred years later, the British physicist J.J. Thomson radically changed everyone's view of the atom as indivisible. He had been studying cathode rays, currents of electricity inside glass tubes similar to those used in old television sets. He speculated that these cathode rays were, in fact, never-before-seen particles which came from inside the atom. These particles were shown to have a negative electrical charge and were eventually named electrons.

On the positive side . . .

But how could an atom, which didn't have an electric charge, contain electrons? The answer came 14 years later when Ernest Rutherford suggested that lurking inside the atom were also positively charged particles which would balance out the negative charge of the electrons. These eventually became known as protons.

In 1932, it was realized that there were still more objects hiding inside the atom, objects that were similar to protons except that they had no electrical charge.

Today's view

So after two thousand years of being indivisible, atoms now consisted of a central area – the nucleus – containing a number of protons and neutrons orbited by a cloud of electrons. An experiment where a piece of thin gold foil was bombarded with helium nuclei showed that the nucleus was very compact and that the electrons orbited a long way out. The atom was mostly empty space. This led to the famous illustration used to show an atom or radioactivity.

BELOW 3-D model of an atom.

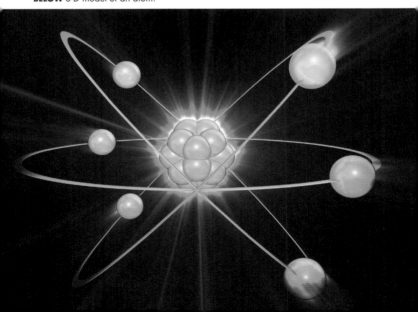

Some equipment which can see atoms and molecules

Field-ion microscope: *This was the device which was used to produce the first ever image of an individual atom in 1955. In this set-up, a sharp needle of a metal, such as tungsten, is placed inside a vacuum chamber and a high voltage applied to it. Any gas still remaining in the vacuum chamber gets strongly repelled by the tip in a direction perpendicular to the needle's surface. This gas can be detected and an intricate image of the tip created.*

Scanning electron microscope: *In this device, a metal electrode – often made from tungsten – is heated up until it produces a beam of electrons. These are accelerated towards another electrode and focussed into a beam just a few nanometers across. The sample you want to study is scanned using this beam and the electrons, which are deflected back from the surface, are picked up by a detector to create an image of the sample being scanned. The images produced are highly detailed and show clear three-dimensional structure.*

Scanning tunneling microscope: *This uses the concept of quantum tunneling to work. A very sharp metallic tip, again often made from tungsten, is brought close to the surface of a sample you want to study. A voltage is set up between the tip and the sample. In this set up, normal physics would argue that electrons could not jump the gap between the tip and the sample surface. However, in the world of quantum physics – the world of the exceedingly small – the laws of physics change and electrons can jump the barrier. The flow of electrons generates an electrical current, which varies according to the bumps caused by the individual atoms of the surface, and these variations are used to create an image.*

LEFT Atomic resolution through a scanning tunneling microscope.

How do lasers work?

Scientist Niels Bohr said that the new model of the atom was wrong. He showed that the electrons could only be in particular fixed orbits, with well defined amounts of energy. In his model an electron moving from one allowed orbit to another would cause the emission of some electromagnetic radiation, such as an X-ray, with an energy equal to the difference between the electron's initial and final energy levels. Electromagnetic radiation of a fixed energy has a distinctive wavelength (or color for light). So every chemical element has a distinctive spectrum of wavelengths corresponding to the energy levels of its electrons, this is why sodium street lamps are a distinctive yellow.

A clash of photons

In a laser, electromagnetic radiation is pumped into a material causing electrons to jump to a higher energy state; the atom is now said to be stimulated.

Left alone, the atom spontaneously emits a photon of electromagnetic radiation and returns to its original state. But if the stimulated atom is hit by another photon, it will emit yet another photon exactly the same as the one that has just hit it, with exactly the same wavelength. It is what is known as "coherent" with the first photon and this is key to how the laser works.

FACT
"Laser" is an acronym for Light Amplification by Stimulated Emission of Radiation. In industry they are used for many things from cutting steel to inscribing the letters on keyboards.

Laser Power

These photons are bounced back and forth between mirrors, with ever more photons produced each time until being released, resulting in a pulse of laser light – all of exactly the same wavelength and coherent. It is this purity that is the source of the laser's power. The power of a laser is usually given as a total amount of energy or as an intensity, the amount of energy it can deliver over a certain target area over a given period of time.

Powerful lasers

Laser	Power	Intensity	Location	Notes
Diocles	100TW	10ZW/cm²	University of Nebraska–Lincoln, USA	Sapphire laser which can be fired 10 times a second.
Hercules	300TW	20ZW/cm²	University of Michigan, USA	Pulse lasts 30 femtoseconds and can be focused on a spot one hundredth of the width of a human hair. Can be recharged in just 10 seconds.
National Ignition Facility	500TW		Lawrence Livermore National Laboratory, California, USA	Ultraviolet laser using a combination of 192 beams, due to be fully operational in 2009.
Omega EP	1PW	>200EW/cm²	University of Rochester, NY State, USA	Four laser beams focus on a target of around one millimeter in diameter. Pulses of one picosecond are possible.
Vulcan	1PW	1ZW/cm²	Rutherford Appleton Laboratory, UK	X-ray laser built using neodymium glass. Can be focused down to 5 microns. one twentieth of the width of a human hair.

How to split an atom

ABOVE Mushroom cloud resulting from the nuclear explosion over Nagasaki — 9 August, 1945.

All atoms are quite similar – they contain a central nucleus of positively-charged protons and/or neutral neutrons and a cloud of negatively-charged electrons orbiting that nucleus. What makes them distinct in a chemical sense is the number of protons (the atomic number). Hydrogen has an atomic number of 1 while calcium has one of 20. When it comes to splitting the atom, this is handy to know.

A process known as nuclear fission, splitting the atom is as simple as breaking a large atom in two. So how exactly do you split the atom? Hitting it very hard sounds silly, but that's just what you have to do.

A bombardment

The method was discovered by a team of scientists in the 1930s, of whom Otto Hahn went on to win a Nobel Prize. Hahn and his colleagues fired neutrons at some uranium and were surprised to find that after the bombardment, chemical elements other than just uranium were present, including barium. They realized that the uranium had been split into two smaller atoms – one of barium and one of the gas krypton.

Fusion power

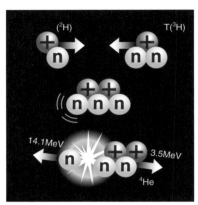

ABOVE Nuclear fusion in action, two small atoms fuse together to make a larger one.

There is another type of nuclear reaction that can release energy too. This is known as nuclear fusion and involves fusing together two small atoms to make a larger one. Fusion reactions are what scientists are certain power the sun and other stars. Unlike fission, where we start off with a large wobbly atom containing loads of protons and neutrons like uranium, we start off with some of the simplest atoms you can imagine.

A simple example

The element hydrogen is the first in the periodic table of elements because of its simplicity – the nucleus consists of just a single proton and a single orbiting electron.

All chemical elements have isotopes. These act in the same way in chemical reactions as the original element but have a different number of neutrons in the nucleus. Hydrogen has two isotopes known as deuterium (which has a proton and a neutron in the nucleus) and tritium (which has a proton and two neutrons in the nucleus).

The theory

In fusion, you strip out the electrons of the elements and then try to force the two nuclei together. There's just one problem. Since both nuclei are positively charged they tend to repel each other, in a similar way that two similar magnets repel each other if you try to force them together.

Use enough force however and you can make them fuse together as shown left.

The result is a nucleus containing two protons and two neutrons, which just happens to be the nucleus of a helium atom. We have fused two nuclei of one atom to make the nucleus of another. As with fission, the by-products of the reaction include both spare neutrons and energy.

Outer space into a bomb

Although we know fusion happens inside stars, getting it to work on Earth is much harder. You need lots of energy to force the original nuclei together and a way of keeping them confined. So far, scientists have had only limited success in creating controlled fusion reactions. The only place where they have succeeded is inside certain types of nuclear weapon, where a bomb which works through fission sets off a subsequent bomb that works using fusion – the results are barely controllable. Perhaps more powerful lasers, which can be used to collapse a pellet containing the necessary elements, will let scientists make a breakthrough.

ABOVE The notorious Fat Man, this is the atomic bomb that was detonated by the United States over Nagasaki, Japan, in World War II.

The biggest nuclear tests

The first ever nuclear test – called Trinity – took place in New Mexico in the US, using plutonium, and its explosion gave a

The power of nuclear devices

Test/device	Date of explosion	Where
Tsar Bomba (RDS–220, Big Ivan)	October 30, 1961	Mityushikha Bay, Novaya Zemlya, Russia
Test 219	December 24, 1962	Novaya Zemlya, Russia
Test 147	August 5, 1962	Novaya Zemlya, Russia
Test 174	September 27, 1962	Novaya Zemlya, Russia
Test 173	September 25, 1962	Novaya Zemlya, Russia
Castle-Bravo (Shrimp, TX21)	February 28, 1954	Bikini Atoll, Marshall Islands
Castle Yankee (Runt II, TX24)	May 5, 1954	Bikini Atoll, Marshall Islands
Test 123	October 23, 1961	Novaya Zemlya, Russia
Castle Romeo (TX17)	March 27, 1954	Bikini Atoll, Marshall Islands
Ivy Mike (Sausage)	November 1, 1952	Enewetak Atoll, Marshall Islands

yield of 20 kilotons. The biggest nuclear explosions on Earth took place during the Cold War, when relations between the US and the Soviet Union were far from friendly and the two superpowers were involved in a deadly arms race.

Yield (megatons)	Notes
58	Detonated at 13,100 feet (4,000 m) above the ground, mushroom cloud reached 40 miles (64 km) high, blast wave was tracked three times around the earth.
24.2	One of the very last Soviet atmospheric tests carried out. Device exploded at 12,300 feet (3,750 m) above ground.
21.1	Device exploded at 11,800 feet (3,600 m) above ground.
20	Device exploded at 12,800 feet (3,900 m) above ground.
19.1	Device exploded at 13,500 feet (4,090 m) above ground.
15	Yield far exceeded predicted 6 megatons because of the tritium bonus, a boost caused by the use of the metal tritium in the design. The larger blast meant many people were unexpectedly affected by fallout, including residents of the other Marshall Islands.
13.5	Half of the yield came from fission of the uranium in the device while the rest came from fusion of partially enriched lithium fuel. The device of 19.8 tons (18 tonnes), nearly 19 feet (5.8 m) long and 5 feet (1.52 m) in diameter, was detonated from a barge.
12.5	Dropped from a plane and exploded at 11,500 feet (3,500 m) above the ground.
11	The device, detonated from a barge, gave a far higher yield than the expected 4 megatons. The mushroom cloud of this test is used on the cover of Megadeth's Greatest Hits album.
10.4	First true hydrogen bomb, where a primary fission bomb causes a secondary fusion bomb to explode. The device, which weighed 82 tons (74 tonnes) and was the size of a building, was not designed as a weapon but more as a test of the concept.

FACT
The detonation of the Tsar Bomba nuclear device lasted just 39 nanoseconds. The power of the explosion caused was about 1 percent of the power of the sun.

How nuclear chain reactions work

When elements undergo fission, you get a couple of smaller atoms, some energy and crucially, a number of neutrons. It is these neutrons that are vital for sustaining a chain reaction. If you remember, fission can be initiated by firing a neutron at an element such as uranium. If the neutrons that result from the fission can then be made to initiate further fission, the process can be self-sustaining.

The critical trigger

Getting a chain reaction to work means ensuring there is enough fissile material around to be triggered by the neutrons from the first reaction. The minimum amount of fissile material that is capable of starting a self-sustaining chain reaction is known as the critical mass.

In nuclear weapons, two amounts of material, both less than the critical mass, are sometimes brought together into one place, pushing the material over the critical mass limit, producing energy, initiating a chain reaction and setting off the bomb.

Particles and waves

Throw a pebble into a flat pond and what do you see?
Ripples that spread out from the spot where the pebble
entered the water of course. What has that got to do with
the world of the really small? Well, you can either think of
the water involved in those ripples as a wave or as a big
collection of water droplets. Treating the water as a wave
would allow you to calculate how long the ripples would
take to reach the pond's edge, while treating them as a
collection of drops might let you calculate how much water
was in the entire pond.

The microscope world is very similar – you can
sometimes treat things as though they are waves and other
times like they are particles. Light is a very good example.

Young's experiment

An experiment carried out in the early 1800s seemed to
show that light acted like the ripples on a pond. The scientist
Thomas Young shone a beam of light onto a sheet
containing two thin slits. Rather than just seeing an image of
two illuminated slits, Young observed a series of alternating
dark and light bands. He explained this by saying that the
light passing through each slit acted just like a pebble
dropping into a pond, causing a series of ripples spreading
out from that point. The ripples from the two slits interact
with each other. Where two peaks in the ripples coincide,
they make a bigger peak – a brighter light. Where a peak
and a trough coincide, they cancel each other out – a dark
spot. Young's experiment appeared to prove that light acted
like a wave.

Einstein's explanation

However, another experiment, involving something called the photoelectric effect, appeared to say the opposite. In this, you shine a light onto a certain type of metal. When the light reaches a certain frequency, electrons start to be expelled from the metal. This can only be explained, said Einstein, if the light is made up of particles.

ABOVE Albert Einstein, his Theory of Relativity would, unknowingly, aid the development of the atomic bomb.

The BIG tiny question

So which is right? Is light made up of waves or particles? It can be considered as both and is one of the freakiest things about quantum mechanics, the laws of physics that apply at atomic and subatomic levels.

Even more bizarrely, other things, such as electrons and protons, can act like they are both particles and waves some of the time.

The forces of nature

Scientists now believe that everything in the universe interacts with everything else through one or more of four fundamental forces. This means that the invisible string that causes the Moon to rotate around the Earth, the weird effect where two similar magnets try to push each other apart, the glue keeping atoms from flying apart, and anything else where one object causes another to start moving, speed up or slow down is down, to one of four basic interactions.

Gravity: *This is the stuff that causes apples to fall out of trees, that makes the planets circle the sun and the force that traps light inside a black hole. It has an unlimited range, meaning that there is a gravitational attraction between even small chunks of rock at opposite ends of the Universe. It is typically the weakest of all the four forces but dominates at large distances where the strength of the other forces is considerably lessened.*

Electromagnetism: *This is the force that stops you pushing two similar magnets together and is also the force that is behind the operation of the electric motor. Like gravity, it has an unlimited range but is vastly stronger on microscopic scales. Two electrons might experience an electromagnetic force 10^{36} (i.e. 1 with 36 zeroes after it) times stronger than the gravitational force between them.*

Strong force: *This is what keeps the atomic nucleus from flying apart. It remained unobserved for most of history because of the extremely short range over which it operates – typically 10^{-15} meters, the size of a proton or neutron. However on these scales, it is by far the strongest force of all, 100 times more powerful than electromagnetism.*

Weak force: *This force only has an impact on very small scales – less than 10^{-18} meters, smaller even than a proton. Although it is called the weak force, it is only weak in comparison to the strong and electromagnetic forces. It is the weak force that causes radioactive beta decay.*

How particle accelerators work

Somewhere in a beautiful part of the world . . .

You might be surprised to know that 330 feet (100 m) below the beautiful French countryside near the Swiss border sits what is claimed to be the world's largest machine – the Large Hadron Collider (LHC). It is set within a 16 mile (26 km) long tunnel, containing what will be the world's most powerful particle accelerator when it switches on in 2008.

A quick introduction to particle acceleration

What is a particle accelerator? The particles in LHC's case are protons – positively charged particles. LHC's particles are accelerated using powerful superconducting electromagnets to speeds approaching the speed of light. Getting the protons to collide involves sending bunches of particles rotating around the experiment in opposite directions. The way the LHC works means that these interact at four points around the circumference of the collider.

ABOVE The ATLAS detector searches for new discoveries in the head-on collisions of protons, learning about the basic forces that have shaped our universe.

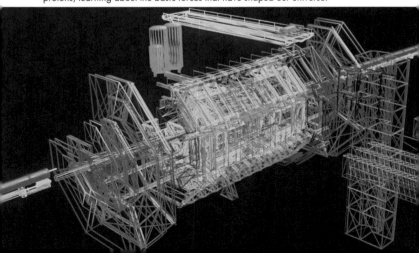

Linear accelerators, such as SLAC, accelerate particles in a straight line using electrified plates towards a target.

The crucial point is that the ultra-high energy from the collision can be used to create a shower of new particles. Albert Einstein's $E = mc^2$ says matter and energy are interchangeable. Sophisticated detectors record everything that is created at the point of collision and sometimes there are new, exotic things which have never been seen before.

The world's biggest particle experiments

ATLAS	LHC, CERN, Geneva, Switzerland	Atlas, one of the experiments at the LHC, is the largest particle physics experiment ever. The detector measures 150 feet (46 m) long by 82 feet (25 m) wide by 82 feet (25 m) high. Some 2,100 physicists from 37 countries will work on it when it fires up in 2008. It hopes to discover why particles have mass, and the nature of dark matter and energy, the constituent of 96 percent of the universe.
BaBar	Stanford Linear Accelerator Center, USA	Electrons and positrons are smashed into each other to create exotic particles known as B mesons. The experiment aims to discover why there is far more matter than anti-matter in the universe.
CDF	Fermilab, USA	Experiment carried out on Tevatron, the world's most powerful accelerator until LHC opens. It smashes protons and anti-protons together. The top quark was discovered here.
STAR	Relativistic Heavy Ion Collider (RHIC), Brookhaven National Laboratory, USA	The detector is as big as a house and weighs 1,200 tons (1,900 tonnes). RHIC smashes together the nuclei of atoms such as gold and the STAR scientists watch what happens. Investigating the early moments after the Big Bang.
Super-KamiokaNDE	Mozumi mine, near Hida, Japan	This experiment is a huge chamber fuel of 50,000 tons (35,400 tonnes) of water surrounded by detectors that can pick up flashes of light caused by the passage of neutrino particles. The experiment is looking to see whether the proton, thought to be a totally stable particle, ever breaks down into other particles.

Fundamental Particles

Scientists now believe that everything around us is made up of basic building blocks called fundamental particles – so-called because scientists believe that they are not made up of smaller constituent particles, see the table below.

There are two classes of particle – quarks to the left, and leptons to the right. The difference between the two classes is that quarks are affected by something called the strong force – the force which keeps atoms from flying apart – while leptons are not affected by it.

The fundamental particles are also arranged in three generations. Particles in each generation are heavier than those in earlier generations. The particles in the second and third generations are inherently unstable and quickly convert into particles in the first generation through various processes. Therefore the vast majority of the ordinary matter we are able to observe is made up of the particles in the first generation.

Up and down quarks don't exist on their own, but are always combined with other quarks to make up larger particles. The proton inside every atomic nucleus, is made up of two up quarks and one down quark, bound together by the strong force mentioned before. But neutrons are made up of two down quarks and one up quark.

	Quarks		Leptons	
First generation	Up	Down	Electron (e-)	Electron neutrino (ne)
Second generation	Charm	Strange	Muon (m-)	Muon neutrino (nm)
Third generation	Top	Bottom	Tau (t-)	Tau neutrino (nt)

All about antimatter

One of the freakiest things in the world of the very small is antimatter. It's regularly mentioned in science fiction novels and movies but it is very much a reality. It seems that every particle we mentioned in the table of fundamental particles on the previous page has an antimatter counterpart, so we have an anti-up quark, an anti-tau neutrino and so on. The anti-electron even has its own name – the positron. It is this anti-particle that was the first to be detected and the one which has been studied more than any other.

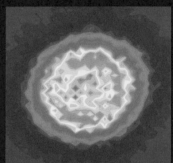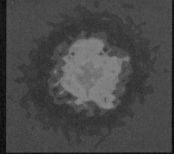

ABOVE Particles and anti-particles annihilate one another, producing photons.

Opposites

Anti-particles are the mirror images of their corresponding particle. The positron is positively charged while the electron is negatively charged. Other weirder, fundamental characteristics of these particles are also equal and opposite. Physical laws such as the conservation of energy and momentum mean that particles and their anti-particles can destroy each other if they meet – their opposing properties just cancel each other out. This is known as annihilation. These annihilations produce energy.

Einstein's famous E = mc² equation tells us that energy and mass (the scientists' name for weight) are interchangeable. The combined mass of a colliding electron and the positron can be converted into energy (in the form of radiation such as X-rays for example) in the blink of an eye. Bizarrely, the reverse can also happen. Energy can suddenly be transformed into mass as long as the laws of conservation hold. This means you could create an electron and positron just from pure energy. In fact, scientists believe this is just what is happening around us all the time – we just don't notice it.

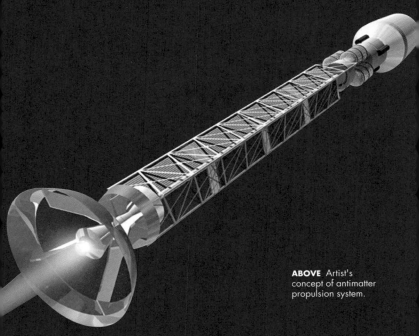

ABOVE Artist's concept of antimatter propulsion system.

THE VERY BIG

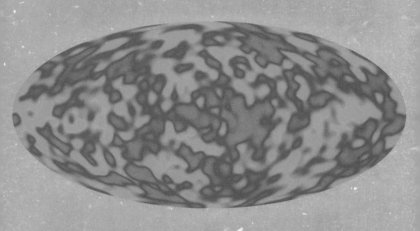

Measuring cosmic distance

Stars are a very long way away. Our Sun is the closest and is still at 93 million miles (150 million km). The next one, Proxima Centauri, is a whopping 25 trillion miles (4×10^{13} km) away. So how do we measure such distances?

Parallax

One method is to measure something called parallax. To get an idea of what this means, hold a finger up in front of you with a distant object behind it, then close first one eye and then the other while looking at the finger. You'll notice that your finger appears to move relative to the distant object. This apparent movement can be used to measure the distance to your finger and the same technique can be applied to measure the distance to the stars. Think about how the Earth moves in its orbit around the Sun over the course of a year. If you look at where Proxima Centauri is in December in relationship to its neighbors and then look again in June, you will notice a small difference in its position. This is because we are looking at it from two points

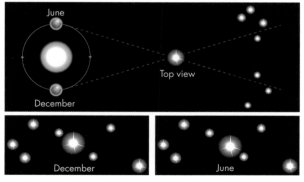

ABOVE Measuring distances beyond the Solar System.

in the Earth's orbit that are 186 million miles (300 million km) apart (twice the distance to the Sun). If you measure the angle that the star seems to have moved, simple trigonometry will tell you the distance.

Unfortunately, because they are so far away, this method can be used for only a limited number of stars. The most advanced parallax measuring equipment, onboard a satellite called Hipparcos, can only use this for stars up to 1,600 light years away – a distance still inside our own galaxy.

Another way of measuring distance

Although very little seems to change in the night sky, the brightness of some stars is constantly changing. Known as variable stars, their brightness can vary dramatically over as little as a few hours or as long as a few weeks.

One type of variable star – a Cepheid variable – varies its brightness over a very regular cycle. Rather usefully, the amount of time it takes a Cepheid to vary is directly related to the star's average brightness.

Why is this useful? Well, if we have two Cepheids which vary over the same time scale and we know the distance to one of them – perhaps through the parallax method – we can then use the difference in how bright they appear relative to each other to measure the distance to the second.

FACT

If someone stands at the bottom of your garden holding up a candle and you measure how bright it looks, then that same person runs off somewhere far away and you measure how bright the candle looks then; if you know the distance to the bottom of your garden, you can measure the second distance based on the difference in brightness. This is an example of a Cepheid variable – often known as a standard candle.

How stars work

In a "normal" star like the Sun it is the nuclear fusion reactions of elements like hydrogen and helium at their core which provide the outward pressure necessary to stop the star from collapsing under gravity. But what happens when the nuclear fuel runs out?

1. *When the hydrogen at a star's center runs out, fusion stops and the core pressure no longer supports it against gravitational collapse.*

2. *This core collapse causes the central temperature to rise, eventually rising high enough to allow the fusion of helium into carbon and oxygen, which requires more energy than hydrogen fusion. Although hydrogen has run out in the core, hydrogen may still be present in the outer layers of the star and fusion of this continues, causing the outer layers to expand outwards.*

3. *When the helium in the core runs out, the process repeats itself – the core collapses while fusion of helium in the outer layers continues.*

4. *The next element which may then undergo fusion is carbon. If the star contains sufficient material, this chain of fusion continues through a series of chemical elements until the star has a core of fusion-produced iron. Iron is a special element because you can no longer gain extra energy from nuclear fusion involving it. As a result, the star's energy source is gradually exhausted as the fusion in the outer layers uses up all of those elements. Stars end their lives in different ways depending on how large they are. There are three main star types.*

LEFT Collapse can take many billions of years.

Red giants

Our Sun and stars, with a mass of between 0.5 and 5 times that of the sun, eventually turn into red giants as they get older. The red giant stage occurs when helium is being fused at the center of the star and hydrogen is being fused in the extended outer layers.

The Sun will have spent about 10 billion years fusing hydrogen before turning into a red giant, when it will expand to beyond the orbit of the planet Mercury. It will spend around a billion years in this stage.

White dwarfs

In stars that contain eight times the material of the Sun or less, the collapsing core never reaches the point where the temperature is high enough to start the fusion of chemical elements beyond helium. Instead, a core of inert carbon and oxygen builds up and the outer layers of the star billow out into space, forming a planetary nebula.

ABOVE A red giant exhausts the supply of hydrogen in its core and it fuses the hydrogen into an outer shell.

Since there is no fusion going on in the core, it begins to collapse again. If the matter remaining in the core (remember that the outer layers have been blown away) is less than 1.38 times the mass of the sun it will collapse until something called electron degeneracy pressure – an unwillingness of electrons to be forced close together – stops it. The core is still quite hot and continues to glow, it is now known as a white dwarf. Over time, it will cool down and eventually become a black dwarf.

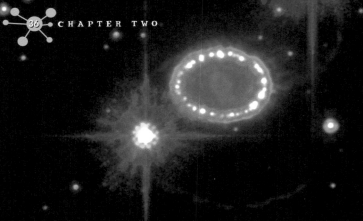

ABOVE A supernova is a stellar explosion that can briefly outshine a whole galaxy.

Neutron star

For stars containing more than eight times the mass of the Sun, the fusion process continues all the way until nickel is produced, which eventually turns into iron. With no pressure holding it up, the star begins to collapse. There is so much material that the electron degeneracy pressure is not enough to stop the collapse and the atoms are split into their constituent particles – protons, neutrons, and electrons. The great energy forces together the protons and electrons to create more neutrons with the release of vast numbers of neutrinos. The unwillingness of neutrons to be forced together stops the collapse and a shock wave travels outwards through the star, with the outer layers flung off in a supernova. The neutron core that remains is known as a neutron star.

Black holes

For really massive stars, there is a different end to their life. Even the unwillingness of neutrons to be pushed together cannot withstand the collapse of some stars and after a supernova has thrown off the outer layers of the star, the remaining core material keeps collapsing until it is so small and dense that nothing can escape its gravity, even light.

How stars get their names

ABOVE Sirius A and B.

Many of the brightest stars in the sky have specific names. Some of the best known are Sirius (the Dog Star), the brightest star in the sky, Polaris (the Pole Star), a star which you can use to find your bearings, and Betelgeuse, if only because people like pronouncing it beetle juice.

Lots of stars have names that were given to them by Arabic astronomers. Betelgeuse is one but there are many that begin with "al" ("the"). Altair means the eagle, Algol means the ghoul.

Another method of naming stars is to base it on the constellation they are in along with a Greek letter to denote their brightness within that constellation. The brightest star is known as alpha (a), the next as beta (b) and so on. The well-known star Alpha Centauri is therefore the brightest star in the constellation of Centaurus (the centaur).

The problem facing astronomers is that as telescopes have got better, the stars they are able to observe are ever more faint. There are now so many that it's impossible to come up with proper names at all, and most stars are just known by a jumble of letters and numbers.

One survey of the sky, the Sloan Digital Sky Survey (SDSS), carried out in 2006 revealed 100 million objects, and that only covered a quarter of the night sky. Stars that were found in the survey are known by codes such as SDSS J073910.48+333353.8. The numbers indicate the star's position, using special coordinates employed by astronomers.

The constellations

When you look up at the sky, it's easy to see why ancient peoples saw patterns among the stars there. These patterns come in two types: constellations and asterisms. These days, the sky is divided into 88 official constellations. Their names reflect what the ancients believed could be drawn in the sky if you joined the stars by lines. Some of the constellations are

Northern sky

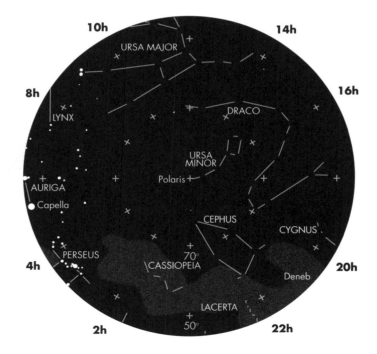

very well known, particularly the constellations of the zodiac, which you will see in horoscopes – Cancer, Capricorn, Aries. Others are well known too – Orion the hunter, for example.

Asterisms are recognizable collections of stars with their own names either within or shared between a number of constellations. Two examples are the Big Dipper – the seven brightest stars in the constellation of Ursa Major; and the Summer Triangle – made up of Altair, Deneb and Vega.

Southern sky

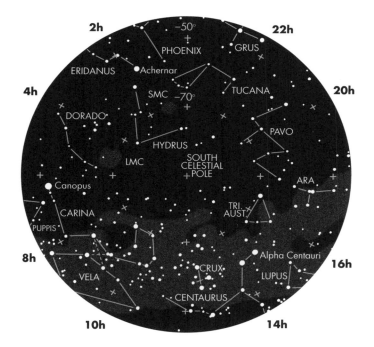

Naming your own stars

Another, unofficial, way that stars get their names is through "star registries" which let you choose your own name for a star on payment of a fee. You then get a certificate showing the new name. The international astronomers association IAU does not recognize the names, and says: "Like true love and many other of the best things in human life, the beauty of the night sky is not for sale, but is free for all to enjoy."

Despite not being official, thousands of people have named stars, calling them some rather unusual names below:

Top five weird names people have given stars

1. **Little Miss Boobsy:** *This star is in the constellation of Pegasus, the winged horse. You'll need a good telescope to be able to spot it though as it's around three million times fainter than Sirius.*

2. **Lord Truffle Pig:** *This yellowy-white star is about one and a half times the size of the Sun. It is in the constellation of Cancer the crab and lies about 685 light years from the Earth.*

ABOVE Truffle-hunting pigs.

3. Porcupine Lobster: *This orange-colored star, in the constellation of Ursa Major (the Great Bear), is around 2,150 light years away from Earth and is likely to be just a little smaller than our Sun.*

4. Sexy Minx: *We can presume that whoever Sexy Minx is, she (or perhaps he) is a Pisces, because that's the constellation where this orange star lies. With a surface temperature of around 4,000 degrees, Sexy Minx is hot, hot, hot.*

5. Smelly-Butt-Josh: *Named by a long-suffering sister for her seemingly odorsome brother, this white star is in the constellation of Sagittarius and will be about two or three times bigger than the Sun. It is visible in a good pair of binoculars.*

Some odd constellations

The air pump, Antilla: *Created in the 18th century by the French abbot and astronomer Nicolas Louis de Lacaille, it contains very few bright stars. The constellation was named in honour of the invention of the air pump by the Frenchman Denis Papin. Unfortunately, the air pump was actually invented by a German, Otto von Guericke.*

Berenice's hair, Coma Berenices: *This used to be part of neighboring Leo, where it was thought to represent a tuft of hair on the lion's tail. Named after Queen Berenice II of Egypt who, according to legend, cut off her long hair as an offering to the gods if her husband returned home safely from war. He did return and the gods (or perhaps some wily astronomer) placed the hair in the night sky.*

The right angle, Norma: *Another Nicolas Louis de Lacaille creation. It has no really bright stars – its two brightest were "stolen" and handed to the neighboring constellation of Scorpio. It does include some impressive deep-sky objects, including the open cluster NGC 6087 and the Fine-Ring Nebula.*

The generator, Machina Electrica: *Now no longer with us, this constellation was dreamed up by the German astronomer Johann Bode. He created it out of part of the constellation of Cetus but it was never popular and the name disappeared. Thankfully.*

What is outer space made of?

Space is a very good description for what exists between the stars and galaxies – there really isn't very much of anything around. In fact, there is so little stuff around that there's probably only a single atom in every cubic centimeter of space – a vacuum better than ever created on Earth.

Galaxies

Galaxies, collections of millions and billions of stars, are thought to have formed from minuscule variations in the gas and dust thrown out from the Big Bang. They come in a variety of shapes and sizes.

Spiral galaxies

Spiral is the "classic" galaxy look – around 70 percent of all the galaxies observed so far are of this type. Stir a cup of coffee and add some cream and you get the idea. With most you get a central blob surrounded by a disk of dust and gas in which there are usually two tightly wound arms. Although it's hard to imagine, our own galaxy – the Milky Way – is also a spiral galaxy. From Earth, we see the Milky Way as a bright band of stars stretching across the sky. This is because the Sun and the Solar System are within the galaxy's disk.

Barred spirals

Spiral galaxies come in two sorts – ordinary spirals, like our near neighbor the Andromeda Galaxy – and barred spirals, with about half belonging to each category. In a barred spiral there is a central blob with a thick line or bar of stars, gas and dust which passes through it. The spiral arms then wind off the end of this bar. The bar is thought to be a channel for gas and dust which feeds the birth of new stars. There is strong evidence that the Milky Way is a barred spiral.

Elliptical galaxies

Most of the remaining galaxies are elliptical – they have the shape of an ellipse (oval). Some of these galaxies are nearly circular while others are elongated ovals, although this may be down to how we see them from Earth rather than any significant difference in their shape – imagine how different a frisbee looks depending on the angle you see it. Many astronomers now believe that elliptical galaxies are the result of a collision between two spiral galaxies in which the gravitational interactions between them smoothes out any spiral structure.

Irregular galaxies

Some galaxies don't quite fit into any categories and have some very unusual shapes, these account for perhaps a quarter of all galaxies that we know. Astronomers believe the odd shapes are because the galaxies are being distorted by gravity, perhaps through collision or near miss by a galaxy in close proximity.

BELOW Sombrero galaxies are sustained by a supermassive black hole.

Big Bang

Most cosmologists agree that the Universe was created 13.7 billion years ago in the "Big Bang." At that time, all that we see in the Universe – the gas and dust in the stars and galaxies; the rocks, water and animals we see on Earth – was crammed into a infinitely small point called a singularity. It is thought that some microscopic fluctuation caused it to become unstable, triggering a rapid expansion.

Expanding space

The name Big Bang was given to this theory by Fred Hoyle, a British cosmologist who actually believed a different theory of the Universe. He called his rival theory the Big Bang as a joke but the name stuck. In fact, the Big Bang is not a Bang at all but a sudden and rapid expansion. Many people imagine there was a huge explosion which sent gas and dust flying out in all directions. This gas and dust would then become stars and galaxies. But this is not the case. Instead, it is the fabric of spacetime itself which is expanding. Think of a balloon inflating. If you draw some dots on the surface of the balloon to represent galaxies, and then inflate it, the dots move further apart.

How do we know?

How do we know the Big Bang took place 13.7 billion years ago? In the 1920s, the astronomer Edwin Hubble (for whom the Hubble Space Telescope is named) used a telescope at the Mount Wilson observatory to look at things known as nebulae (Latin for fuzzy spots). He used Cepheid variable stars to measure how far these blobs were away from Earth to realize the blobs were outside the Milky Way. In fact, they were other galaxies. He then measured how much their light was red-shifted to work out how fast they were moving. He quickly realized that the further away a galaxy was, the faster it was moving away from us.

Hubble's Law

Drawing a graph of the speed against distance showed it was a straight line. If you think of how you know that a car traveling at a constant 30 miles (50 km) an hour which is 90 miles (145 km) away has been traveling for three hours, you'll realize you can work out how long the galaxies have been traveling – how long it has been since the Big Bang. He got the numbers slightly wrong but we have now worked back to realize that it was 13.7 billion years ago.

ABOVE The Hubble Space Telescope has resolved some long standing mysteries, helping to form the theory that the expansion of the universe is accelerating.

How big is the Universe?

You may think that if the Universe is 13.7 billion years old, and the fastest thing in the Universe travels at the speed of light then the Universe is 13.7 billion light years across? Wrong. In actuality, because it is space itself which is expanding – so not limited by the speed of light – the Universe could be any size. In fact, cosmologists now think that the Universe is around 156 billion light years across.

The first 380,000 years of the Universe

Cosmologists are keen to find out what happened at the very beginning after the Big Bang. Things changed very rapidly in the first three minutes after the Universe was born.

Time after Big Bang	What was happening
10^{-43} seconds	Early on, there were not four fundamental forces of the Universe but just one superforce that later shattered into four.
10^{-43} to 10^{-36} seconds	The strong force separates from the others.
10^{-36} to 10^{-32} seconds	The fledgling Universe expands rapidly through something called inflation, which smoothes and flattens it out. The size of the Universe grows rapidly in this time – its volume expands 10^{78} times – that's 1 with 78 zeroes after it.
10^{-32} to 10^{-12} seconds	The weak and electromagnetic forces separate out.
10^{-12} to 10^{-6} seconds	The first quarks form.
10^{-6} seconds to 1 second	The first protons and neutrons and their anti-matter counterparts form but quickly annihilate each other.
1 second to 3 minutes	The first electrons and positrons (antimatter counterpart to the electron) form but quickly annihilate each other.
3 minutes to 380,000 years	The Universe is full of radiation from the annihilations, in the form of photons. At the end of this era it is the photons that are seen today as the cosmic microwave background.

Other cosmic theories

One other idea says the Universe has lasted forever but is expanding, with new material created in bursts every now and again to maintain the overall density of the Universe; another – the Mixmaster Theory – sees the Universe oscillating between the shape of a pancake and a cigar.

Some big telescopes

Large Binocular Telescope: *Built at a cost of US$120 million, the LBT has an unusual design, using two 27.5 feet (8.4 m) wide mirrors mounted next to each other, like a pair of binoculars. Working together, the two mirrors have the power of a single mirror of 38.7 feet (11.8 m). The 600-ton (545-tonne) telescope sits on the summit of Mount Graham in south-eastern Arizona and made its first binocular observation, of the galaxy NGC 2770, in January 2008.*

Gran Telescopo Canarias: *This US$200 million telescope made its first observations in July 2007. It has a main mirror divided into 36 hexagonal segments with the equivalent light-gathering power of a circular mirror 34.1 feet (10.4 m) wide, and sits on top of a dormant volcano on La Palma, Canary Islands. Its aim is to study planets orbiting stars other than the Sun.*

Keck 1 and 2: *The Keck observatory also sits on top of a dormant volcano – Mauna Kea, in Hawaii. The observatory actually has two telescopes, each with 32.8 feet (10 m) wide mirrors and weighing 300 tons (275 tonnes) apiece. The first began observations in 1993 with the second following three years later. The telescopes study the formation of galaxies and look at so-called gravitational lenses, celestial objects that bend light from other, more distant, objects.*

Hubble Space Telescope: *It may not be as big as the telescopes above, but the Hubble telescope has one huge advantage. Its 7.9 feet (2.4 m) wide mirror is outside the Earth's atmosphere, orbiting our planet every 97 minutes, meaning that the images are very clear and have given astronomers some stunning views of the Universe. Expected to operate until 2013, to be replaced by the bigger telescope, the James Watt.*

FACT
No man-made object, let alone a camera, has traveled far enough into the reaches of space to be able to take an external photograph of the Milky Way galaxy.

Red and blue shift

You'll see a good example of science in action every time a police car with its siren blaring passes you in the street. Everyone knows that the sound changes as it passes you. It's not that the siren changes its tune but it is down to something known as the Doppler effect. This is where the sound waves coming from moving objects get squashed or stretched depending on whether something is moving toward or away from you. Something similar happens with light waves (as well as other electromagnetic radiation such as X-rays and microwaves). Things moving away from you get redder (in scientific-speak their wavelength gets longer) while things moving toward you get bluer (the wavelength gets shorter). As a result, this effect is known as red or blue shift.

You can't normally notice this effect with light because the object has to be moving at huge speeds but you can with distant galaxies.

Cosmic latte

In 2001, Karl Glazebrook and Ivan Baldry of Johns Hopkins University did a study looking at the color of light from a variety of galaxies in the Universe. Their research showed that the light in the Universe had an average color of beigy-white. In an article in the *Washington Post*, Glazebrook jokingly asked for suggestion for a name for the color. One reader suggested the term cosmic latte. The name stuck.

ABOVE The color of space, a cosmic latte.

Cosmic microwave background

Back in the 1960s, two engineers were trying to come up with a new communications device that used microwaves – yes those things that you use to zap meals in microwave ovens. The problem was that the receiving dishes they were using to collect the microwaves kept picking up lots of noise. They checked everything, even looking at the dishes to make sure no birds had left any deposits on them. In the end, they realized that the noisy microwaves were coming from the sky, and not just from one point, but everywhere they turned the dish. With a leap of imagination, they realized that the noise was, in fact, the leftover radiation from the Big Bang, now cooled down to a chilly three degrees above absolute zero.

What do we see when we look at the sky?

We now know that the speed of light is not infinite but instead travels at 186,000 miles per second. This has one very hard-to-grasp consequence. When you look up at the night sky, you are seeing stars that lie at a range of enormous distances from us.

Looking into the past . . .

Proxima Centauri, our nearest star other than the Sun, is just over four light years away. So when you look at it, the light you are seeing has taken four years to reach your eyes. The star Sirius, however, is eight light years from us so the light from Sirius has taken eight years to reach us. This means that we are seeing Proxima Centauri as it was four years ago and Sirius as it was eight years ago. Either star could have exploded or turned bright pink since then and we wouldn't know. So when we look in the night sky, we are looking into the past and for very distant objects, we are looking very far into the past. And if we look at very distant objects, we can look back to around the time of the Big Bang.

ABOVE The Crab nebula, in the constellation of Taurus, was first discovered in 1731.

Top five coolest objects you can see in Google Sky

Google's view of the skies (www.google.com/sky) is a wonder of the Internet; a zoomable map of the Universe with stunning images.

1. Cartwheel Galaxy: *The unusual wheel shape is down to a collision with another galaxy that initiated the birth of vast numbers of new stars. X-rays are emitted from the rim from black holes, eating up gas and dust.*

2. Centaurus A: *Around 11 million light years from Earth, gases in the galaxy are now emitting high-energy X-rays, heating up a vast ring of material surrounding it.*

3. Crab Nebula: *Known to astronomers as NGC 1952, this is the remnant of a supernova that was visible for two years after it exploded in the year 1054.*

4. Sombrero Galaxy: *This spiral galaxy in the constellation of Virgo is viewed side on from Earth. There's a huge glowing bulge of stars bisected by a dark line of dust, making it look like the hat in the name.*

5. The Pillars of Creation: *These are the clearest images of where stars are born in stellar nurseries. These enormous pillars of gas and dust are part of the Eagle nebula, around 7,000 light years from Earth.*

Black hole oddities

Black holes are among the weirdest objects in the Universe. We have never seen one directly but astronomers believe they lurk at the center of most galaxies. Because they are so massive, they have a number of weird effects on things that pass close by or fall inside them.

Slow friends

Time, as well as space, is distorted around heavy objects such as black holes. Imagine that an adventurous friend wants to visit the inside of a black hole and you watch his journey with your powerful telescope. Due to the time distortion, your friend seems to travel slower and slower as he approaches the black hole until he appears to stop completely. He hasn't really stopped, as far as your friend is concerned his speed hasn't changed. This is one of the weird effects of "relativity."

Spaghettification

The strength of gravity decreases the further you are from the center of an object. Gravity is slightly weaker at your head than at your toes. Near a black hole, things are very different. If you were falling feet first into a black hole, the gravity at your feet would be much larger than at your head. If you weren't killed by all the radiation present around a black hole, the thinner and thinner stretching of your body – like spaghetti – would literally rip you apart.

ABOVE Spaghettification in action.

White holes and wormholes

The equations of Einstein's general theory of relativity tell us that black holes can exist. But they also tell us that other strange objects can exist too – white holes and wormholes.

White holes are the opposite of black holes – material pours out of them, rather than being attracted in. However, they have never been observed, although scientists believe they might exist. One tantalizing possibility is of a black hole in one part of the Universe and a white hole in a completely different part, connected by something called a wormhole, a bridge between two distant locations. Sadly, it seems that we couldn't use them to travel huge distances in the blink of an eye. The radiation and the enormous gravitational forces would smash us and our spaceships to smithereens.

Gravitational lenses

Albert Einstein was the first to realize that really strong gravity could bend light. Prior to that, scientists thought that because light doesn't have mass, like an apple falling out of a tree for example, then it wouldn't be affected. Einstein says that space itself gets distorted by heavy objects such as stars. Think of a sheet with a watermelon on it. The dip in the sheet around the watermelon represents how space is bent near a heavy object. Roll a grape close by the watermelon and its path gets bent. This is what happens to light too.

Now imagine you have something really heavy, like a galaxy containing millions of stars. Some way beyond it you have another galaxy. If you look from the Earth at the first galaxy, the light from the galaxy beyond gets bent around the near one. What it looks like from Earth is one galaxy surrounded by arcs of light, the distorted image of the far galaxy. This effect is known as gravitational lensing.

Pulsars, quasars and other weird stuff

As well as all the trillions of normal stars out in the Universe there are some other oddities that astronomers have come across in their searches of the night sky.

Pulsars

In the 1960s, astronomers observed an object which seemed to be acting like some sort of cosmic lighthouse, sending out a beam of radiation towards the earth which was extremely regular in its timing. They initially thought it was a signal being sent by aliens but now know it is a fast spinning neutron star with a huge magnetic field. This means it emits huge beams of gamma rays, with pulsars that emit X-rays. The beams sweep round as quickly as a few milliseconds up to a few seconds.

ABOVE Pulsars, a cosmic lighthouse.

Quasars

The name quasar is a shortened version of quasi-stellar radio source.

Like pulsars they emit huge amounts of radiation but as radio waves rather than X- or gamma rays. Initially thought to be stars or something similar they are now thought to be the central cores of galaxies that lie huge distances away from us and are only visible as featureless blobs. It is thought that supermassive black holes at the centers of these galaxies are the cause of the emission of the radio waves.

The most distant objects ever

The record for most distant object ever seen is broken all the time but the current holder is a galaxy called A1689-zD1, thought to be around 13 billion light years away from us (therefore we are now seeing it as it was 13 billion years ago, or just 700 million years after the Big Bang). It's impossible to see in visible light but the Hubble Space Telescope and the Spitzer Space Telescope teamed up to look at the object in infrared. The galaxy is thought to be undergoing a massive burst of star formation.

BELOW Bright young thing, A169-zD1 (pictured in the frame below) was formed just 700 million years after the Big Bang.

SPACE STUFF

How to remember the order of the planets

You never know when learning the order of the planets might come in handy. Most people use a mnemonic – a sentence whose words start with the same letters as the things you are trying to remember. Until 2006, you could have used My Very Excellent Mother Just Sent Us Nine Pizzas (Mercury, Venus, Earth, Mars, Jupiter, Saturn, Uranus, Neptune, and Pluto). That was until the governing body of the world's astronomers, the International Astronomical Union, decided that Pluto was really just a dwarf planet. And two other heavenly bodies – Ceres and Eris – were promoted to the same status. A National Geographic competition was held to find a new mnemonic and 10-year-old Maryn Smith from Montana, came up with: My Very Exciting Magic Carpet Just Sailed Under Nine Palace Elephants.

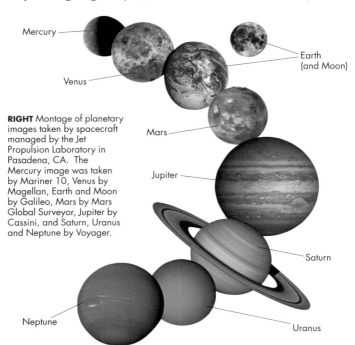

Mercury

Earth
(and Moon)

Venus

RIGHT Montage of planetary images taken by spacecraft managed by the Jet Propulsion Laboratory in Pasadena, CA. The Mercury image was taken by Mariner 10, Venus by Magellan, Earth and Moon by Galileo, Mars by Mars Global Surveyor, Jupiter by Cassini, and Saturn, Uranus and Neptune by Voyager.

Mars

Jupiter

Saturn

Neptune

Uranus

Cool things about planets

Mercury: *Mercury is about two fifths the diameter of Earth and whips around the Sun once every 88 Earth days. Unlike the Earth, which rotates on its own axis once every 24 hours, Mercury takes 59 Earth days to spin. Being so close to the Sun, Mercury has the greatest variation in temperature on its surface between day and night. Just before dawn, the temperature is a chilly –292° F (–180° C). By the afternoon, the temperature has soared to 806° F (430° C), hot enough to melt tin.*

Venus: *If you thought Mercury was hot, just take a look at Venus. Despite being further from the Sun, the temperature on the surface reaches more than 496° F (480° C). The reason is the thick clouds of carbon dioxide that cloak the planet and drive the temperature up, just like a greenhouse. The atmosphere also contains clouds of sulfuric acid, although when it rains, the acid evaporates by the time it reaches the surface of the planet.*

Mars: *Mars has around half the radius and a tenth the mass of the Earth. Most of the planet's atmosphere is made up of carbon dioxide (95 per cent). Mars is known as the red planet, thanks to the red and pink dust and rocks that cover its surface. A Martian volcano, Olympus Mons, is the largest known in the Solar System. It is 16 miles (27 km) high from base to summit, making it three times higher than Everest, and is 340 miles (550 km) across.*

Ceres: *The dwarf planet Ceres, named after the Greek goddess of the harvest, was discovered at the beginning of the 19th century. Astronomers had long expected there to be a planet between Mars and Jupiter, largely because of a discredited physical law that seemed to dictate the distance of the planets according to a regular mathematical series. However, it proved to be just the first of more than a hundred thousand objects between the two neighboring planets, now called the asteroid belt. Ceres, the largest of these objects, was reclassified as a dwarf planet in 2006.*

Jupiter: *This is the largest of the planets in our solar system. In fact, it is more than twice as massive as all the other planets put together. Planets like Jupiter are known as gas giants (to distinguish them from the rocky,* terrestrial planets like Earth) but in fact, they are not just made up of gas. The atmosphere is made up of hydrogen and helium, but the pressure and temperature inside the planet is enough to force these elements, which we know as gases, into liquids and even a solid state. The planet's most obvious feature is the Great Red Spot, a giant storm in the planet's atmosphere that has lasted for more than 400 years. Three planet Earths could fit inside the Spot.

ABOVE The rings of Saturn.

Saturn: Saturn is another gas giant. Like Jupiter, it is mainly composed of hydrogen and helium, much of it in liquid form, but has an icy, rocky core, thought to be about 10 to 20 times as massive as the Earth. Observations of the gaseous surface show that the atmospheres at the equator and at the poles are rotating at different speeds, giving rise to super hurricane winds that blow at more than 1,100 miles (1,770 km) per hour. Saturn is best known for its rings – huge discs of small particles of dust and ice that orbit the planet. The rings extend 75,000 miles (120,700 km) from Saturn's surface but are perhaps just 60 feet (18 m) thick.

Uranus: When Uranus was first spotted, by Britain's first Astronomer Royal John Flamsteed in 1690, he mistook it for a star. It was only in 1781 that it was positively identified as a planet by William Herschel. He originally named it Georgium Sidus (Georgian planet) in honour of Britain's King George III but it was later renamed after the Greek god of the sky (rather than a Roman god like the other planets). Unlike Jupiter and Saturn, Uranus is largely made up of different kinds of ice made from water, methane, and ammonia. In 1977, it was discovered that Uranus has its own system of rings, like those of Saturn only much less spectacular.

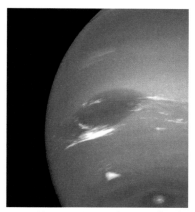

ABOVE Neptune, 17 times the mass of Earth.

Neptune: *Named after the Roman god of the sea, Neptune was discovered in the mid-19th century after its position was predicted, based on oddities in the orbit of neighboring Uranus caused by the disturbing tug of its gravity. It is similar to Uranus, having a rocky core, then a thick icy body and an atmosphere made mainly of hydrogen and helium. The presence of methane means the atmosphere absorbs red light, giving Neptune a vivid blue color. Winds of 2,400 km/h (1,500 mph) blast through the atmosphere, a record wind speed for the Solar System. Vast storms, like the red spot of Jupiter, are seen to tear across the planet's 'surface'.*

Pluto: *Poor old Pluto – demoted from its planetary status in 2006. It is tiny, having 1/500th the mass of the Earth and a fifth of its radius. This means that its density is considerably lower than that of Earth and suggests that it is made up of a mixture of rock and ice. Unlike the other planets, which have nearly circular orbits, Pluto's is more oval. This has the odd effect that it is sometimes closer to the Sun than Neptune. Because Pluto is so far out, the Sun appears like a dot in the sky. This distance also means it's very cold – about 40 degrees above absolute zero.*

Eris: *Eris, or 2003 UB313 as it was formerly known, was discovered by astronomers in 2003, and is the most distant object ever discovered in orbit around the Sun. Originally called the new object Xena, after television's fantasy adventure show, it was officially named Eris, after the Greek goddess of warfare, in 2006. It was the discovery of Eris – 27 percent bigger than Pluto – which led to both of them being designated dwarf planets, along with Ceres. At just under 10 billion miles (16 billion km) from the Sun, Eris is three times further out on average than Pluto.*

Earth's second moon

Did you know that Earth has (at least) two moons? One we know well but the second is a very strange critter indeed. Called Cruithne (pronounced croo-ee-nya), it is just three miles (4.8 km) wide and was only discovered in 1986. The oddest thing about Cruithne (or asteroid 3753) is its orbit. It follows a normal elliptical orbit around the Sun (which means it's not really a moon in the astronomical sense) but the orbit has been synchronized with that of Earth. From our planet, it appears to follow a horseshoe shaped orbit around our planet, making a complete revolution once every 770 years.

What are planets made from?

For a long time people believed that the Moon might be made out of cheese thanks to its Swiss cheese-like pock-marked face. We now know, thanks to the US Apollo and Soviet Luna missions which returned samples from the Moon's surface to Earth, that the surface is made out of very old rock. The Moon is unusual in that we have actually been there, collected samples and brought them back to Earth.

ABOVE Lunar dairy.

Rocks from Mars

Believe it or not, we have rocks that originated on the planet Mars here on Earth. If we haven't been there, how did they get here? Well, scientists believe that rocks were blasted out of the Martian surface by collisions in the past and the rocks have been drifting around in space ever since. When one of these rocks happens to cross the Earth's orbit, they start to burn up in the atmosphere, and we see them as shooting stars. Some of the Martian rocks are so big that they reach the Earth's surface without completely burning up, but this is very rare. NASA says that there are just 34 known Martian meteorites. Virtually all of them contain a mineral called shergottite, after the village of Shergotty in India, where the first Mars meteorite of this kind was found, in 1865. These days, most of the Martian rocks are found in Antarctica or the world's deserts. The dark space rocks are easy to see in the paler and snowy landscapes.

Scientists can be fairly sure that the rocks come from Mars for two reasons. When they date them, they find that they are less than 1.5 billion years old – which makes them relatively young! – and the masses of the elements that make them up, such as oxygen, are found in ratios not seen on Earth or Moon rocks.

FACT

Scientists are seriously considering the possibility that there is life on Mars. In 1996, NASA scientists announced that there was evidence of tiny fossilized bacteria in the make up of rock ALH84001, found in Antarctic ice but originally from Mars. Since then, this rock has been analyzed and academic opinion remains divided. While scientists are confident that the Mars rock contains ancient life, they are still not willing to say that it comes from Mars – it might be Earthly contamination.

The color of chemical elements

Astronomers think that Jupiter's outer atmosphere is mainly hydrogen gas with a bit of helium thrown in. But how do they know this to be the case? It's all down to a technique called spectroscopy. When you heat up or pass electricity through a gas, it gives off a characteristic color. This is why sodium street lamps glow yellow, for example. It is caused by the electrons in the chemical element gaining some energy from the heat and then losing that energy by emitting some light. Different elements give off light of many different colors.

Now think of what happens to the light from the distant Sun as it hits the atmosphere of Jupiter. Some of it is reflected and bounces off in the direction of Earth, which is the light we see when we peer through our telescopes. But light of that special color we mentioned above is absorbed by electrons and then re-emitted and not just in the direction of Earth but in all possible directions. This means that the light we see when we look at Jupiter has less of that special color in it. Astronomers are able to measure this and knowing which chemicals are related to which colors allows them to determine the chemicals that are present in the atmosphere.

FACT
The planets are all different colors when viewed through a telescope. The colors depend on what the atmosphere of the planet is made from. The colors are: Mercury, light gray; Venus, yellow; Earth, blue, brown, green; Mars, red; Jupiter, yellow; Saturn, yellow; Uranus, green; Neptune, blue.

Some cool spaceship propulsion technologies

Rocket engines: *The most widely used method of launching and powering a spacecraft. A fuel, in the form of a solid, liquid, or gas is introduced into a chamber along with something called an oxidizer. The reaction of the two generates heat, causing the mixture to expand. This is forced out of a shaped nozzle at high speed.*

Ion thrusters: *These use fuels such as xenon, mercury, or bismuth. The fuel is ionized (electrons are stripped off) and accelerated in an electric field. The positive nuclei of the fuel are ejected from the engine, re-neutralized with electrons and this pushes the spacecraft forward.*

Mass driver: *This theoretical method of launching a rocket uses a series of electromagnets in a line. The spacecraft sits in a reusable launcher which is accelerated along a rail by firing up the electromagnets in turn. The rocket launches at the end of the rail and the launcher is returned to the start of the rail for re-use.*

Solar sail: *Similar in concept to the sail on a boat. Instead of the wind pushing the vessel along, a solar sail uses the pressure of radiation from the Sun. To make use of this, the spacecraft has a large, foldaway fan of mirrors. The gentle pressure on the mirror pushes the craft along.*

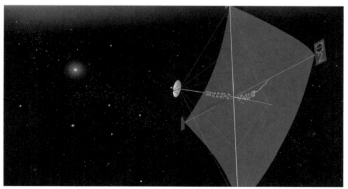

LEFT Solar sails make use of the sun's radiation for propulsion.

Top five space missions

1. Apollo: *Perhaps the most famous series of missions of all, with the goal of landing humans on the surface of the Moon, which it achieved with the arrival of Neil Armstrong and Edwin "Buzz" Aldrin in Apollo 11 on the lunar surface (with Mike Collins orbiting in the Command Module) on July 21, 1969.*

2. Giotto: *One of five in an armada of spacecraft which flew by Halley's Comet on its nearest approach to Earth in 1986. It flew within 375 miles (600 km) of the heart of the comet, taking remarkable pictures of the peanut-shaped 10 mile (16 km) long nucleus. Despite being hit by high-speed particles and spinning off like a top, it was restabilized and used to visit another comet in 1992.*

3. SpaceShipOne: *This was the name of the spacecraft which won the US$10 million Ansari X prize, a prize awarded to the first private team to build and launch a spacecraft capable of carrying three people to 60 miles (100 km) above the Earth's surface, twice within two weeks. Aerospace designer Burt Rutan with financial help from Microsoft's Paul Allen achieved the feat in October 2004.*

4. Vostok 1: *On April 12, 1961, Soviet cosmonaut Yuri Gagarin became the first person to visit space on board this spacecraft. He was away from Earth for just an hour and 48 minutes, reaching an altitude of around 200 miles (322 km) above the surface and traveling at a maximum speed of 17,660 miles (28,421 km) per hour.*

5. Voyager: *The twin Voyager spacecraft were launched in 1977 and their main goal was to study the planets of Jupiter and Saturn, which they examined in incredible detail, revealing active volcanoes on Jupiter's moon Io, for example. Voyager 2 went on to fly by Uranus and Neptune and is the only mission to have visited these planets. Both spacecraft are still operating and giving scientists an indication of what lies beyond the planets.*

Spacecraft orbits and itineraries

If you look up at the stars, your eyes are often drawn to small dots that travel across the sky quickly and smoothly but without the flash of a shooting star. Some of these dots are high-flying jet planes but some, more excitingly, are spacecraft that are orbiting the earth, hundreds of miles above its surface. The glow is caused by the reflection of sunlight off their metallic surfaces.

Geostationary orbit

Arthur C. Clarke, the writer who wrote many great science fiction stories and novels, including the one that went on to become the famous movie *2001: A Space Odyssey*, was one of the first people to recognize the value of geostationary orbits, through an article in the magazine Wireless World in 1945. A satellite in geostationary orbit is one where the satellite keeps the same position relative to a point on the Earth's surface. The satellite isn't actually stationary but is rotating at exactly the same speed as the surface of the Earth, making it seem to stand still in the sky. To achieve this, the satellite has to be in a very particular orbit – at 22,240 miles (35,1791 km) above the Earth's surface – and it needs to be directly above the equator. The orbit, which is sometimes called the Clarke orbit in honor of the writer, is often used for communications and television satellites. Satellites that are at the end of their useful life are often boosted out of this orbit into a graveyard orbit to avoid congestion.

The slingshot effect

Launching a spacecraft from Earth requires an enormous amount of energy so that it can escape the clutches of our planet's gravity. In fact, you need enough power to give an object a speed of at least 7 miles (11 km) per second to escape the Earth. But some spacecraft blast off with even higher speeds. NASA's New Horizons spacecraft, which headed off to Pluto from Earth in January 2006, sped off at 10 miles (16 km) per second, the highest escape speed ever. Even at these speeds, getting anywhere really far away will take too long. That's why missions to the outer planets use the slingshot effect. In this the spacecraft passes close to a planet or moon and picks up speed thanks to its gravity. If it is traveling faster than the escape velocity of the planet, then it won't start to orbit but get flung off at a faster speed. The Voyager 2 probe, for example, used a chance alignment of the planets to gain its current speed of around 35,000 miles (56,300 km) per hour. It is around 6.5 billion miles (10.5 billon km) away from us now.

ABOVE The trajectory of the Voyager 2 fly-by of Jupiter toward Saturn.

Odd stuff in space

Disks

ABOVE Gold disk.

When the two Voyager spacecraft were launched in 1977, they each carried a 12-inch (30-cm) gold-plated copper disk holding audio and picture messages intended for the eyes and ears of alien lifeforms (whatever form that might be).

Ashes

Many people choose to have their cremated ashes scattered at sea, in parks, and so on, but what if space is your favorite place? No problem. Your remains can join those of Star Trek creator Gene Roddenberry, actor James Doohan (Scotty in the original series), and the writer Timothy Leary in space.

Junk

What happens to satellites when they die? Some of them are large enough to survive the intense heat of re-entry and crash down to Earth. Others remain in their orbits, gradually falling apart and slowly falling back to Earth. Many of these bits will eventually burn up in the atmosphere, but until then they pose a risk to working satellites and Shuttle missions. Some of the pieces may be tiny, but they are traveling at high speed – a small fleck of paint once gouged a half inch (120 mm) hole in the window of the Shuttle.

BELOW Distribution of space junk around the Earth.

As a result, NASA's Orbital Debris Program monitors this space junk. It estimates there are 17,000 objects larger than 4 inches (10 cm), 200,000 objects measuring between $\frac{1}{5}$ in and 4 in (1–10 cm) and tens of millions of smaller particles. Larger objects are routinely tracked.

SETI

The Search for Extraterrestrial Intelligence (SETI) was founded as a non-profit organization in 1984. These days, SETI employs 150 scientists. However, the organization also has a lot of help. Its SETI@Home project means you can join in too. Download a piece of software to your computer and when you're not using it, your spare processing power will be used along with those of 3 million other people to analyze observations of the Universe made by radio telescopes to look for signs of intelligent life.

BELOW Is there life beyond Earth?

Some places where we might find extraterrestral life

There is only one place in the entire universe of billions upon billions of galaxies, stars, and planets where we are certain there is life – here on Earth. But where else might we look for life?

Little green men

Science fiction writers and moviemakers have been obsessed with the idea of life on Mars ever since author H.G. Wells terrified readers with his book *The War of the Worlds*. The reason that so many people think there might be life out there is that Mars is very similar to the Earth. It is rocky like Earth, has polar ice caps and has a radius of about half that of our planet. It is widely held that liquid water is necessary to sustain life – Mars has water, but it's all frozen solid.

It might have been different in the past when the planet was warmer, extreme forms of life – such as micro-organisms that can withstand low temperatures, high pressures, and environments that are toxic (to humans) – may well exist.

ABOVE Methane, the gas in cow farts, is an importnt compound in the creation of life.

Do cows fart in outer space?

In March 2008, the astronomers making observations using the Hubble Space Telescope announced they had found a tantalizing clue to the possibility of life on another planet. They detected evidence of methane (the gas found in cow farts) in the atmosphere of a planet called HD 189733b, which is the size of Jupiter. The planet orbits a star 63 light years away in the constellation of Vulpecula.

The discovery is important because methane is one of the compounds thought to be important in the creation of life. Life as we know it is unlikely on the planet as the atmosphere is a sweltering 3092° Farenheit (1,700° C).

The man in the Moon

Finding life on the Earth's moon seems incredibly unlikely as it has virtually no atmosphere, extremes of temperature, and shows very obvious evidence (those craters) that it has been bombarded with high-speed objects from outer space. But it might be that we'd be more likely to find life on the moons of Jupiter. Jupiter has dozens of moons, including four large ones discovered by the Italian astronomer Galileo. Modern-day astrobiologists think that of these, the moon Europa offers the best chance for hosting life. On first glance, it doesn't look appealing – it has a thick, icy surface. Yet astronomers believe that there may be oceans containing water beneath the ice and life may thrive here just as it does under Antarctica's ice.

Possible alien lifeforms

Life on Earth is based on the chemicals carbon, hydrogen, oxygen, and nitrogen. Added to that, biochemical processes require the presence of water. But what if there are alien lifeforms that work on a different chemistry?

Floaters: *The astronomer Carl Sagan came up with the idea that there might be floating aliens in the atmospheres of planets like Jupiter. They might resemble vast jellyfish but be ammonia- rather than water-based.*

Silicon-based lifeforms: *If you look at the periodic table of chemical elements (see page 80), you'll notice that underneath the symbol C (for carbon) is the symbol Si (for silicon). Elements in the same columns have similar chemical properties so some scientists have wondered whether life based on silicon might be possible, Silicon is one of the main ingredients in rock and sand, so imagine living rocks. . .*

Artificial gravity and weightlessness

It was Sir Isaac Newton who worked out exactly how gravity works with his Universal Law of Gravitation. His mathematical formula showed that the strength of gravity depends on three things – the mass of the two objects and the square of the distance between them. What his equation shows is that someone twice as heavy as you would experience twice as much gravity as you. It also means that the strength of gravity between two spheres one meter apart is nine times stronger than if they were three meters apart (since three squared is nine). Put together, this tells us that the gravity felt on different planets varies depending on their mass and size. The mass of the Earth is 6 quadrillion kilograms – around 86 times the mass of the Moon (70 thousand trillion kilograms). However, the Moon's radius is around 3.6 times smaller than that of the Earth. This means that a person standing on the surface of the Moon experiences gravity which is 86 divided by 3.6

ABOVE Astronauts experiencing weightlessness in the Nasa KC-135.

squared (13.4) – around 6.4 times weaker than on the surface of the Earth. This is why film of the Apollo astronauts shows them bouncing around the Moon with apparently little effort.

Weightlessness

Sir Richard Branson is set to launch the first commercial flights to the edge of outer space in 2010. There are currently more than 65,000 people on the waiting list for the expeditions, with flights estimated to cost somewhere in the region of U.S.$100,000. The appeal for many will be the chance to experience weightlessness – sometimes incorrectly known as zero gravity. Newton's law of gravity shows us that we never really escape the clutches of gravity. In fact, weightlessness is another phenomenon altogether. When you stand on the surface of the Earth, what you experience as weight is actually the resistance of the ground pushing upwards on the soles of your feet. In an orbiting spacecraft, the floor is subject to the same gravity as you so it is effectively falling towards the Earth the same amount as you – giving the feeling of weightlessness.

FACT

Every year the moon moves about 3.8cm further away from the Earth. This is caused by tidal effects. Consequently, the earth is slowing in rotation by about 0.002 seconds per day per century. Scientists do not know how the moon was created, but the generally accepted theory suggests that a large Mars sized object hit the earth causing the Moon to splinter off.

QUIRKY
CHEMISTRY

Earth, air, fire, water and other crazy ideas

For a long time after the Ancient Greeks, scientists believed that everything you can see or feel in the world was made of a combination of four things: earth, air, fire and water. Everything else that you can't see or feel was classed as idea, void, space or quintessence depending on your culture. These classifications gave things their coldness, hotness, wetness or dryness according to how much of each of the four elements was present.

We now know that everything we see is made up of chemical elements. These elements are made from subatomic particles such as protons, neutrons and electrons. And protons and neutrons are in turn made up of even more minute particles, called quarks.

The states of matter

Ice cubes, water and steam are just three different forms of the same thing, but at different temperatures – the first is a solid, the second a liquid and the third a gas. Scientists call these the states of matter and all chemical compounds can exist in all of these states, given the right combination of temperature and/or pressure. This is odd to imagine since we experience the world within a narrow temperature range in which, say, helium is always a gas. Even salt can exist in all three of these forms – heated to 1472° F (800° C) it melts; heated to 2669° F (1465 ° C) it turns into a gas.

ABOVE Gas.

ABOVE Liquid.

ABOVE Solid.

Odd states of matter

As well as the familiar three states of matter, other bizarre forms exist.

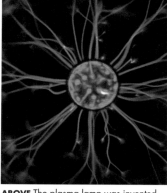

Plasma: *Sometimes called the fourth state of matter. In a plasma, the negatively charged electrons of all the chemical elements present have been stripped off, leaving a soup of electrons and the positively charged atomic nuclei.*

ABOVE The plasma lamp was invented by Nikola Tesla in 1894.

Superfluid: *A liquid which behaves in an odd way because of some quirk of its composition. For example, some types of liquid helium can creep up the side of the container they're in.*

Supercooled liquid: *A liquid which has been cooled down in a special manner to below its natural freezing point. In a laboratory, liquid water can be cooled down to at least −104° F (40° C) without turning into ice.*

Superheated liquid: *A liquid which has been heated to more than its boiling point without turning into a gas. Superheating can sometimes happen when you heat water in a microwave oven. But if you disturb the liquid, by adding coffee granules perhaps, some of the liquid can instantly flash into gas, causing a rapid burst of vigorous bubbles (potentially scalding you).*

Thixotropic liquid: *A liquid whose thickness changes over time, or depending on how hard it is shaken or squeezed. Tomato ketchup is thixotropic, which means it often gets stuck in the bottle until you give it a hard shake.*

ABOVE Shake it to wake it.

Bizarre chemical reactions

Here are some of the freakiest chemical reactions you can see in the lab. Definitely don't try them at home!

The Briggs-Rauscher reaction: *In this fascinating reaction, you mix together three colorless liquids. Almost immediately, the mixture turns amber, then after a while it changes to blue-black before then turning colorless once again. This strange color cycle then repeats itself. The three liquids are solutions of hydrogen peroxide, potassium iodate, sulfuric acid, malonic acid, manganese sulphate and starch.*

Gummy bears and potassium chlorate: *Potassium chlorate is a chemical that is often used in fireworks and in some disinfectants. If you heat it up and then add a gummy bear, the sucrose in the candy starts to react violently with the potassium chlorate, producing clouds of oxygen, bright light and plenty of noise.*

Baking soda and vinegar: *Take a balloon and put some baking soda into it. Then get a cylinder and fill it with vinegar. Carefully put the balloon over the mouth of the cylinder without spilling any of the baking soda. Hold the cylinder round the middle and then hold up the balloon so the baking soda falls into the vinegar. Amazingly, the balloon inflates and the cylinder becomes colder as heat is lost during the reaction.*

Sodium and hydrochloric acid: *This is one of the wow effects that many chemistry teachers demonstrate to their pupils. A small amount of hydrochloric acid is put in the bottom of a test tube. With careful handling, a tiny piece of sodium metal is dropped into the acid. The sodium bobs on the surface of the acid, generating a huge flame and a small explosion. The reaction also generates lots of gaseous fumes.*

The Coke-Mentos experiment

This dramatic experiment was first popularized in 2002 by the science teacher and magician Steve Spangler. You take a handful of Mentos candies and drop them in a large plastic bottle of sweet fizzy soda, such as diet cola. The result is a spouting geyser of fizzy soda.

ABOVE The rapidly expanding carbon dioxide of the Mentos creates a spectacular eruption.

The experiment works with both diet and regular fizzy sodas but diet sodas are "less sticky when you're cleaning it up".

There are numerous explanations for why this happens, although most, including Spangler, believe it is a physcial rather than a chemical reaction.

Bubbles of carbon dioxide in the fizzy soda are held together by a force known as surface tension. The gelatin and gum arabic from the dissolving Mentos reduces this surface tension, causing the bubble to break. The Mentos is pitted with lots of tiny holes, which act as sites for new bubbles to generate. These new bubbles rise rapidly upwards, causing the soda to gush out.

Most dangerous chemicals in the world

In 2001, the United Nations Environment Program issued a list of 12 chemicals which it called the most dangerous in the world. 50 countries agreed to aim to restrict their use as part of something called the Stockholm Convention. This

Chemical	What it kills
Aldrin	A pesticide used to kill termites and grasshoppers. Aldrin-treated rice is thought to have killed hundreds of birds off the Texas coast.
Chlordane	A pesticide that kills termites but also kills shrimps and ducks. It is also thought it might cause cancer in humans.
DDT	A chemical used to protect against the disease malaria. However, DDT remains in the soil for years, stops bird's eggs from developing properly and may accumulate in human breast milk.
Dieldrin	A pesticide used to control textile pests. It is toxic to fish and frogs and it remains in the soil for years after it is sprayed on crops.
Dioxins	Released into the atmosphere when things don't burn properly. The incineration of hospital waste is a key source of these. Dioxins can cause birth defects.
Endrin	An insecticide and a chemical also used to control mice. Lasts in the soil for up to 12 years and is highly toxic to fish.

convention was called in the wake of several high profile chemical disasters costly to both human life and the natural environment. The nasty chemicals are:

Chemical	What it kills
Furans	Chemicals that are a by-product of making printed circuit boards and also found in waste incinerator smoke. They are possible carcinogenic (cancer-causing) chemicals.
Heptachlor	Chemical used to kill grasshoppers and malaria-carrying mosquitoes. Can kill rabbits and other wildlife and may be a carcinogen.
Hexachlorob-enzene (HCB)	Used to treat against fungi. People eating seeds treated with HCB have been known to develop skin problems. Some have even died.
Mirex	Used to kill fire ants and also found on some fire retardant materials. Toxic to some fish and shellfish and may be a possible carcinogen.
Polychlorina-ted biphenyls (PCBs)	Used in the manufacture of electrical transformers and also as additives in paint and plastics. Toxic to fish and, when eaten by humans in food, cause swelling and vomiting and development problems in young children.
Toxaphene	A pesticide widely used on cotton crops in the 1970s. A possible carcinogen and highly toxic to fish. It lasts for up to 12 years in soil.

The periodic table of chemical elements

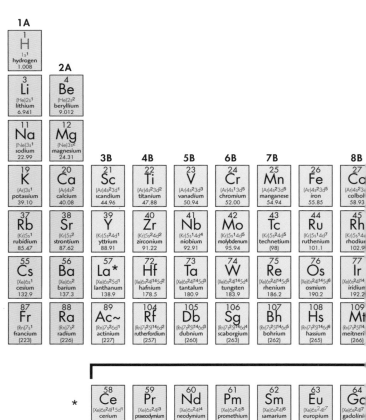

- Element names in outline are liquids at room temperature.
- Element names in gray are gases at room temperature.
- Element names in black are solids at room temperature.
- Elements are color-coded: alkaline metals (blue); transitional metals (aqua); semi-metals and non-metals (purple); noble gases (yellow); lanthanide series (orange); actinide series (green).

		3A	4A	5A	6A	7A	2 He $1s^2$ helium 4.003
		5 B (He)$2s^22p^1$ boron 10.81	6 C (He)$2s^22p^2$ carbon 12.01	7 N (He)$2s^22p^3$ nitrogen 14.01	8 O (He)$2s^22p^4$ oxygen 16.00	9 F (He)$2s^22p^5$ fluorine 19.00	10 Ne (He)$2s^22p^6$ neon 20.18
		13 Al (Ne)$3s^2 3p^1$ aluminium 26.98	14 Si (Ne)$3s^2 3p^2$ silicon 28.09	15 P (Ne)$3s^23p^3$ phosphorus 30.97	16 S (Ne)$3s^23p^4$ sulpher 32.07	17 Cl (Ne)$3s^23p^5$ chlorine 35.45	18 Ar (Ne)$3s^23p^6$ argon 39.95

	11B	12B					
28 Ni)4s²3d⁸ nickel 58.69	29 Cu (Ar)4s¹3d¹⁰ copper 63.55	30 Zn (Ar)4s²3d¹⁰ zinc 65.39	31 Ga (Ar)4s²3d¹⁰4p¹ gallium 69.72	32 Ge (Ar)4s²3d¹⁰4p² germanium 72.58	33 As (Ar)4s²3d¹⁰4p³ arsenic 74.92	34 Se (Ar)4s²3d¹⁰4p⁴ selenium 78.96	35 Br (Ar)4s²3d¹⁰4p⁵ bromine 79.90	36 Kr (Ar)4s²3d¹⁰4p⁶ krypton 83.80
46 Pd r)4d¹⁰ alladium 06.4	47 Ag (Kr)5s¹4d¹⁰ silver 107.9	48 Cd (Kr)5s²4d¹⁰ cadmium 112.4	49 In (Kr)5s²4d¹⁰5p¹ indium 114.8	50 Sn (Kr)5s²4d¹⁰5p² tin 118.7	51 Sb (Kr)5s²4d¹⁰5p³ antimony 121.8	52 Te (Kr)5s²4d¹⁰5p⁴ tellurium 127.6	53 I (Kr)5s²4d¹⁰5p⁵ iodine 126.9	54 Xe (Kr)5s²4d¹⁰5p⁶ xenon 131.3
78 Pt atinum 95.1	79 Au (Xe)6s²4f¹⁴5d¹⁰ gold 197.0	80 Hg (Xe)6s²4f¹⁴5d¹⁰ mercury 200.5	81 Ti (Xe)6s²4f¹⁴5d¹⁰6p¹ thallium 204.4	82 Pb (Xe)6s²4f¹⁴5d¹⁰6p² lead 207.2	83 Bi (Xe)6s²4f¹⁴5d¹⁰6p³ bismuth 208.9	84 Po (Xe)6s²4f¹⁴5d¹⁰6p⁴ polonium (209)	85 At (Xe)6s²4f¹⁴5d¹⁰6p⁵ astatine (210)	86 Rn (Xe)6s²4f¹⁴5d¹⁰6p⁶ radon (222)
110 Ds s²5f¹⁴6d⁹ astadtium 271)	111 Uuu (272)	112 Uub (277)						

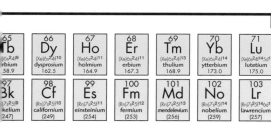

65 Tb)6s²4f⁹ erbium 58.9	66 Dy (Xe)6s²4f¹¹ dysprosium 162.5	67 Ho (Xe)6s²4f¹¹ holmium 164.9	68 Er (Xe)6s²4f¹¹ erbium 167.3	69 Tm (Xe)6s²4f¹³ thulium 168.9	70 Yb (Xe)6s²4f¹⁴ ytterbium 173.0	71 Lu (Xe)6s²4f¹⁴5d¹ lutetium 175.0
97 Bk)7s²5f⁹ kelium 247)	98 Cf (Rn)7s²5f¹⁰ californium (249)	99 Es (Rn)7s²5f¹¹ einsteinium (254)	100 Fm (Rn)7s²5f¹² fermium (253)	101 Md (Rn)7s²5f¹³ mendelevium (256)	102 No (Rn)7s²5f¹⁴ nobelium (259)	103 Lr (Rn)7s²5f¹⁴6d¹ lawrencium (257)

Oddities of the periodic table

The periodic table is a way of organizing all of the chemical elements that exist in the world. Chemical elements in the same column tend to have similar chemical properties because there is an underlying periodic nature to the number of protons and electrons they contain.

As you can see from the picture of the table, these elements are called different things:

Halogens: *These are highly reactive chemicals which means they are rarely found on their own but instead in compounds with other chemicals, such as sodium chloride (common salt). Fluorine is the most reactive element known and can even react with glass so chemists don't keep it in glass bottles. Halogen lamps usually contain a gas like nitrogen plus a tiny amount of a halogen gas.*

Rare earths: *The first of these elements, ytterbium, was first discovered in rocks in Sweden in the late 18th century. They are typically found as oxides (compounds containing oxygen) rather than by themselves. They were called rare because they were thought to be scarce but in fact they are relatively common in the Earth's crust.*

Alkali metals: *This group of elements is highly reactive and, like halogens, are more often found as compounds rather than on their own. In the chemistry lab, they are usually stored in oil to stop them reacting. Sodium and potassium are widespread on earth and are essential for human life. Hydrogen is usually placed at the top of the alkali metals, and, in some extreme circumstances, acts just like one.*

Inert elements: *These elements, sometimes known as the noble gases, don't like combining together with other elements due to the way their electrons are organized. In fact, there are no known stable compounds which include helium or neon – they just exist on their own unless artificially created in the lab.*

How some chemical elements were discovered

Many chemical elements, such as carbon and copper, have been known since ancient times because they occur naturally on the Earth. However, other chemical elements were harder to find.

ABOVE Albertus Magnus belived that stones had occult properties.

Arsenic

Discovered in the 13th century by an alchemist called Albertus Magnus when he heated up a type of orange-yellow stone called orpiment with soap. The reaction between the two created arsenic as a liquid metal.

Phosphorus

Discovered by chemical means (rather than being found in nature), it was uncovered in a very unusual way in the middle of the 17th century. German alchemist Hennig Brand heated up some urine until there was just some fine powder left over. He burned this powder and it produced a bright flame.

Oxygen

On heating a compound containing mercury in 1774, Joseph Priestley found that the gas it produced made a candle burn more brightly. However, because he believed in a theoretical substance called phlogiston he thought he had discovered air minus the phlogiston.

Potassium, sodium, calcium, barium

Found in 1807 and 1808 by Humphrey Davy, He is perhaps best known for inventing the Davy lamp, used by miners. He discovered these four elements by passing electricity from a primitive battery through various types of salt.

How new chemical elements are named

Newly discovered chemical elements, which are all unstable and can only be made for fleeting moments in nuclear laboratories, are given temporary names based on a system agreed by the International Union of Pure and Applied Chemistry. Elements are distinguished and named by their atomic number (the number of protons in the atom's nucleus). The following table is used to create the name,

Digit	Code	Symbol
0	nil	n
1	un	u
2	bi	b
3	tri	t
4	quad	q
5	pent	p
6	hex	h
7	sept	s
8	oct	o
9	enn	e

The highest atomic number so far discovered in a chemical element is 118. However, it is expected that further elements will be found. Therefore, if an element with atomic number 126 is found its name will be constructed from this table with the suffix -ium, as follows.

Un + bi + hex + ium

So, element 126 if it is discovered will be called unbihexium and have the chemical symbol Ubh.

Only once the element's existence has been independently confirmed is it given another permanent name. The element with atomic number 110, discovered in 1994 at a laboratory in Darmstadt, Germany, was originally called ununnilium (with the symbol Uun). Now that its existence has been confirmed, it has been renamed darmstadtium (with the symbol Ds).

Chemical names of everyday things

Table sugar	$C_{12}H_{22}O_{11}$	Table sugar is known among chemists as sucrose and is made up from two other types of sugar joined together – glucose and fructose. It is easily digested by the body, giving an instant energy boost.
Aspirin	$C_9H_8O_4$	The painkiller aspirin, which is also known as acetylsalicylic acid, was first sold by the German pharmaceutical company Bayer in the late 19th century but similar chemicals have been used for medicinal purposes since long before that.
Plastic bottles	$(CH_2=CH_2)_n$	Bottles, and plenty of other objects in the house, are made from a plastic called polyethylene, which is a long chain of smaller compounds called ethenes. The n in the chemical formula means that the bit inside the brackets is repeated again and again.
Chili peppers	$C_{18}H_{27}NO_3$	The substance that makes a chili pepper hot is known as capsaicin or 8-methyl-N-vanillyl-6-nonenamide. Pure capsaicin is 3,000 times hotter than the average jalapeño.
Olive oil	$C_{17}H_{35}NO_3$ $COOH$	Olive oil is actually a complicated mixture of lots of different fats and vitamins but its main constituent is known as oleic acid or cis-9-octadecenoate. It is this compound which is thought to give olive oil its healthy properties.

FACT
The name Hydrogen comes from the Greek words 'Hydro' and 'Gen' which mean 'water generator' and it is the most abundant element in the universe.

Some really strong acids

Acids are one of the most interesting types of chemicals, burning through objects and combining with other chemicals to create explosions, flames and clouds of gas. It is no surprise that mad scientists always seem to be tinkering with them. Not all acids are corrosive, however. The carborane superacid H(CHB11Cl11), which is one million times stronger than sulfuric acid, is entirely non-corrosive, whereas the weak acid hydrofluoric acid (HF) is extremely corrosive and can dissolve, among other things, glass and all metals except iridium.

Acid name	Chemical formula	Strength*	Uses
Fluoroantimonic acid	HSbF6	1019	Plastics manufacture and petroleum industry
Magic acid	FSO3H–SbF5	1018	Plastics manufacture and petroleum industry
Carborane acid	H(CHB11Cl11)	1,000,000	Over-the-counter counter vitamin manufacture
Trifluoromethanes ulfonic acid	CF3SO3H	1,000	Petroleum industry
Fluorosulfuric acid	FSO3H	1,000	Pharmaceutical manufacture

*times stronger than pure sulfuric acid

EXTREME
EXPERIMENTS

Some big pieces of scientific equipment

Atlas: *One of four experiments being carried out on the Large Hadron Collider, due to start running at CERN, Geneva, Switzerland in 2008. The experiment consists of a barrel shaped detector 151 feet (46 m) long, 82 feet (25 m) in diameter. It weighs more than 7,000 t (6,350 metric tonnes) and contains more than 1864 miles (3,000 km) of electric cables, enough to stretch from Chicago to San Francisco. It will be used to probe inside the very smallest objects in the universe and recreate the conditions soon after the Big Bang.*

Jet: *The Joint European Torus is based at the Culham Science Centre in Oxfordshire in the UK. It is the largest nuclear fusion research facility in the world. A torus is shaped like a donut and this contains a plasma of deuterium and tritium, which are basically forms of hydrogen. The plasma is heated up to a hundred million degrees to allow fusion to take place which then generates energy. Unfortunately, the energy required to run Jet is currently more than it generates, although future fusion facilities which build on what scientists have learnt at Jet should do better.*

Team: *The Transmission Electron Aberration-corrected Microscope is the world's most powerful microscope and is based at the National Center for Electron Microscopy at the Lawrence Berkeley National Laboratory in California. It is due to become fully operational in 2009 but is already able to pick out details as small as 0.05 nanometers. Team has been used to pick out individual atoms in crystals of gallium nitride. (Aberration is a focussing problem afflicting existing electron microscopes which causes images to be fuzzy. Team uses a combination of special magnetic lenses to correct this.)*

Some chance scientific discoveries

Animal electricity

In 1781, a professor of anatomy in Bologna, Italy noticed a very strange phenomenon. He was working with the legs of dead frogs and noticed that they contracted when a nearby

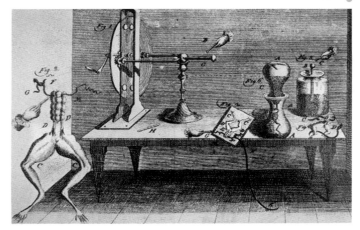

ABOVE Galvani's accidental discovery paved the way for the modern battery.

machine produced electric sparks. The effect was even more spectacular when an assistant placed a metal scalpel on the muscle in the frog's leg. What Galvani had discovered was the role of electricity in the bodies of animals. However, Galvani thought there was some kind of electrical fluid at work in the frog's leg. A few years later, Alessandro Volta (for whom the volt is named) showed what was causing the contraction and went on to develop the voltaic pile.

Vaccination

The French scientist Louis Pasteur was investigating the disease cholera and how it affected chickens. He asked his assistant Charles Chamberland to inject some chickens with some cholera germs to make them ill. However, Chamberland went on holiday and only injected them on his return. Rather than dying, the chickens remained alive, as if the germs had become weaker while Chamberland was away. They then injected the chickens again with some fresh germs and they still remained well. Pasteur realized that the earlier injection had given the chickens immunity against the disease.

Although another scientist, Edward Jenner, had discovered that people who had been exposed to cowpox were protected from the killer disease smallpox, Pasteur was the first to discover that a weakened form of a disease could be injected to protect from the full form. Pasteur called the process vaccination (from the Latin for cow) in honor of Jenner's earlier work.

Penicillin

The bacteriologist Alexander Fleming, when working at St Mary's Hospital in London, noticed that a sample of staphylococcus bacteria had been contaminated by a blue-green mold. At the edges of the mould, the bacteria were being killed off. Fleming grew a sample of the mold and found that it killed off a number of disease-carrying

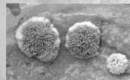

ABOVE *Penicillium notatum.*

bacteria. He found that the mold was one called *Penicillium notatum* and he named the bacteria-killing component of the mold penicillin. It wasn't until the 1940s that scientists found a way to produce high-quality penicillin in bulk for medical purposes.

Microwave oven

The US firm Raytheon was a manufacturer of magnetron tubes used in military radar but a chance discovery led to the development of the microwave oven. The firm's engineer Percy LeBaron Spencer was standing near a magnetron tube, which generates microwave radiation, and noticed that a candy bar in his pocket had melted. He carried out further tests using popcorn and an egg. In 1947, the firm demonstrated the world's first microwave oven, the Radarange, which cost US $2,000 to US $3,000 each, only affordable for commercial uses. Eight years later, the firm developed the first domestic oven, a snip at just US $1,300.

What is an explosion?

If you have watched any Hollywood action blockbuster, you'll have seen a few explosions – terrorists trashing subway trains, action heroes totaling buildings – but what is an explosion? The dictionary defines it as a sudden release of energy, caused by something like a chemical or nuclear reaction, often accompanied by a rapid rise in temperature and the release of large volumes of gas.

Mushroom clouds

These instantly recognizable clouds are most commonly associated with nuclear explosions but can in fact be caused by any big explosion, such as a volcanic eruption or the detonation of a large amount of traditional explosive such as TNT. The explosion causes a vast ball of superhot gas which

ABOVE A mushroom cloud from Redoubt Volcano during an eruption in 1989.

starts to rise because it is less dense than the surrounding air. Dust and debris are pulled upward from the ground, causing the familiar stalk. Eventually, the ball of gas reaches a point where it is of a similar density to the surrounding air and starts spreading outward, forming the cap. Currents of heat can also cause the edges of the cap to curl under.

Implosions

These are the opposite of explosions and happen when an external force causes something to collapse in on itself, concentrating matter and energy. For example, a submarine whose metal structure fails underwater might implode. Nuclear weapons often use implosion too. A series of explosions in a spherical shape are triggered, causing the nuclear fuel inside the sphere to implode. If it reaches critical mass, a secondary explosion occurs, quite often to devastating effect.

FACT

It's hard to believe but custard powder, and other fine powders such as flour and coal dust, can explode. It's all down to surface area. Consider a log fire. You can make a bigger fire by chopping a big log into smaller ones because the surface area increases and more oxygen, which fuels the combustion, can reach the log surface. With a fine powder, each bit of powder is tiny but the surface area of each adds up to make a very large surface area. Introduce a flame or a spark into a cloud of custard powder and it will rapidly go up in flames, as the owners of several custard powder factories around the world have learned to their cost.

Some explosives

TNT: *Trinitrotoluene has become a standard among explosives, thanks to its ease of handling (unexpected knocks will not detonate it) and an ability to be used in many environments. It is yellow in color and was used as a dye when first discovered in 1863. It was not until the 1900s that its potential as an explosive was recognized. The power of other explosives are now commonly compared to the power of TNT.*

Nitroglycerin: *A clear, oily liquid with the advantage of rapid detonation but the huge disadvantage of being highly unstable, meaning that a jolt can cause it to explode. Discovered in 1846 in Italy and widely used in the First World War. Incredibly, the substance is also used medically to treat heart conditions.*

Dynamite: *Invented by Alfred Nobel in 1866, it solved many of the problems of handling nitroglycerin. He added a chalky white rock called kieselgur, which stabilized the formula. The mixture is then packed into cardboard cylinders. The money Nobel made from his invention is the source of funds for the annual Nobel prizes.*

C4: *A type of plastic explosive, in a putty-like form, allowing it to be molded into different shapes. It is 30 percent more powerful than TNT. C4 typically includes a tracer chemical, allowing its origin to be tracked.*

Explosive strength is measured by a test named the Trauzi lead block test, developed by Isidor Trauzl in 1885.

The test is performed by loading a 10-gram foil-wrapped sample of the explosive into a hole drilled into a lead block. The hole is then topped up with sand, and the sample detonated electrically. After detonation, the volume increase of the cavity is measured. The result, given in cm3, is called the Trauzl number of the explosive.

Scientific fraud

The competition between scientists to make new discoveries and win funding is so intense that some resort to faking their results. Because some areas are so specialized it is certain that there are cases of fraud that have gone undetected, though other scientific fraudsters have been found out and held to account.

The clones that weren't

South Korean researcher Hwang Woo-suk published a scientific paper claiming he had cloned 11 humans. Successful cloning techniques, such as those used to produce the sheep clone Dolly, take DNA from a parent and implant it into an egg which then develops normally. Hwang Woo-suk was shown to have falsified his work and resigned from his job. However, investigators analyzing his research have found that he had actually made a ground-breaking discovery in extracting stem cells (the basic building blocks of the human body which can develop into anything the body needs) from unfertilized eggs.

The organic transistor . . . or not

German Jan Hendrik Schön was a talented physicist with a bright future in his field of condensed matter physics and nanotechnology, winning several prizes for his work in his early 30s. He claimed to have discovered a way to make a transistor out of organic materials rather than silicon, amongst other discoveries. When other scientists looked at his work, they realized that some of the data in his scientific papers seemed to be repeated in different experiments. A committee appointed to investigate his work found that he had used fraudulent data in 16 instances, although the co-authors on his papers were unaware. 21 papers co-written by Schön have now been withdrawn.

Odd scientific papers

Scientists publish the results of their experiments as scientific papers, articles in respected magazines that are read by other scientists interested in the same sorts of research. It's hard to get a paper published, which makes it all the more surprising that the following saw the light of day.

Cure for a headache

This paper sounds fairly straightforward until you realize that it looks at the reasons why woodpeckers don't get sore heads when pecking trees. Ivan Schwab of the University of California, Davis found that woodpeckers have particularly bony skulls and cartilage attached to their lower jaw bones to cushion the impact of up to 12,000 pecks in a single day.

Sword swallowing and its side-effects

In 2006, Brian Witcombe, a consultant radiologist and Dan Meyer of the Sword Swallowers' Association International published this paper in the British Medical Journal. They asked 46 sword swallowers about any problems they had after swallowing swords for entertainment purposes. Unsurprisingly,

RIGHT Hard to swallow.

sore throats were common, especially when first starting out as a sword swallower or when swallowing an oddly shaped blade. Injury was also common when the swallower was distracted.

Pressures produced when penguins poop

In late 2003, the scientific journal Polar Biology published this amusingly named paper. It was written by Victor Benno Meyer-Rochow of the University of Oulu, Finland and József Gál of Loránd Eötvös University, Hungary and revealed that penguins generate far higher pressures inside their bodies than humans in order to expel their poop beyond the edges of their nests.

ABOVE Under pressure.

Consequences of erudite vernacular utilized irrespective of necessity

The title of this paper may be confusing but its subtitle – Problems with using long words needlessly – explains what it is about. Daniel M. Oppenheimer of Princeton University noted that most guides to writing style encourage people to use simple words yet undergraduate students often use long words in an effort to seem more intelligent. In fact, his research showed the opposite – that you should use clear and simple words if you want to appear more intelligent.

Five scientific effects you can see around you

Doppler effect: *The changing sound of a police siren, caused by the sound waves being scrunched together when the police car is approaching you and being stretched out when it is moving away, altering the frequency of the sound.*

The different speeds of light and sound: *During a thunderstorm, you see the flash of lightning and then hear the crack of thunder some time later because sound and light have different speeds. By knowing their speeds you can work out how far the storm is away from you.*

Refraction: *If you look at someone standing on the bottom of a swimming pool with half their body in the water and half out, their legs appear to be shorter than they really are. This is because of refraction, the bending of light when it passes between two different materials.*

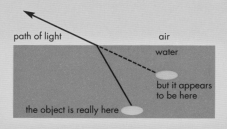

path of light

air

water

but it appears to be here

the object is really here

Pressure: *If you have ever taken a packet of crisps on board a plane you'll know that the bag puffs out as the air inside is at a much higher pressure than in the cabin. Once you open the packet, the pressure is then released (or equalized). The pressure in the cabin of most commercial planes is roughly the same as at the top of a mountain 8,000 feet (2,400 m) high.*

Gravity: *When you fire a catapult your projectile goes straight, then curves down till it hits the ground because gravity is pulling it down.*

If you had a super catapult and fired it from a tall building it would travel further, taking longer to hit the ground. Now, imagine standing on top of a mountain as high as the Moon with a superduper catapult that can fire Moon-sized projectiles at the same speed as the Moon orbits the Earth. Guess what happens – the Moon-sized projectile goes into orbit. The projectile drops as before but the curvature of the Earth means it never hits the ground.

The Philosopher's stone and other things that don't exist

When scientists are seeking a theory to explain the world as we see it, they often come up with ideas that seem to explain the experimental results. Except that often these ideas are fixes that are disproved by later theories.

Phlogiston: *A theoretical substance that all flammable materials were thought to contain that allowed them to burn. It was thought to be a substance or field (like a magnetic one rather than one of corn) which allowed rays of light to travel from one place to another. However, accurate measurements of the speed of light show that it does not exist.*

Philosopher's stone: *A substance that alchemists believed could turn cheap metals into gold. It was believed to be deep red in color and thought to contain a chemical element known as carmot.*

Yellow and black bile: *These substances (along with blood and phlegm) made up the so-called humors of medieval medicine. Any illness was believed to be caused by an excess or lack of one of the four, which could be cured by feeding a patient with various potions, letting them bleed or having blood sucked out by leeches.*

Plum pudding atoms: *When J. J. Thomson discovered the electron in 1897, he came up with an idea that electrons and protons were evenly spaced throughout an atom like a plum pudding. Ernest Rutherford's theory of the atomic nucleus meant plum pudding atoms were ditched.*

FACT
The philosopher's stone was also believed to be an elixir of life, useful for rejuvenation and possibly for achieving immortality. For a long time it was the most sought after goal in Western alchemy.

Freaky experiments

Elephant on acid

In 1962, Louis J. West and Chester M. Pierce of the University of Oklahoma decided to see what the effects of the drug LSD, or acid, would be on an elephant. They fired an LSD dart into the hind quarters of Tusko, a 14-year-old elephant at Oklahoma City Zoo. Shortly afterwards, Tusko trumpeted, collapsed, defecated and had a type of epileptic fit. An hour and 40 minutes later, Tusko was dead.

Glow-in-the-dark cats

In the labs of Gyeongsang National University in South Korea, Professor Kong Il-keun and his team took some skin cells from a Turkish Angola cat, genetically modified them with a virus that caused them to glow in the dark and then transplanted these cells into an egg of a donor cat. Three kittens were born that glowed red in the dark.

ABOVE The glowing cats could help advance our understanding of genetic diseases in humans and assist in the development of stem cell research.

Large Hadron Collider

Some people believe that when the Large Hadron Collider is switched on in 2008, it will create mini black holes because of the high energies involved. These could start devouring everything around them, eventually swallowing the entire Earth. A Hawaiian botanist has even threatened to sue scientists there to stop them turning it on.

The ear mouse

In 1995, Dr Charles Vacanti shocked the world with photos of a mouse with a human ear on its back. Critics said it showed why genetic engineering was all wrong. In fact, the "ear" was a piece of artificially grown cartilage and involved no genetic modification at all.

Falling cats

In 1894, the French Academy of Science wanted to find out why cats land on their feet when falling from height. The answer came when Étienne Jules Marey devised a film camera that took lots of frames in a short time and showed how the cat first twists its front, then rear paws toward the ground.

FACT

The small size, light bone structure, and thick fur of cats decreases their terminal velocity when falling — they may also spread out their body to increase drag and slow even more.

TRICKY TIME

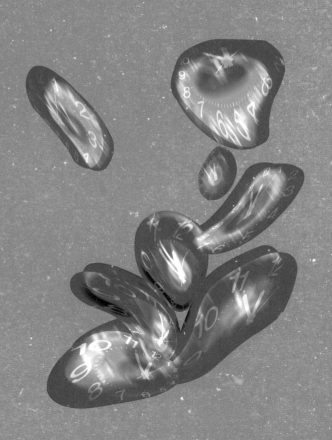

The speed of light

Light, which is a type of electromagnetic radiation in the same way as radio waves, X-rays and microwaves, travels at a fixed speed in a vacuum of around 186,000 miles (300,000 km) per second. It is usually represented by the symbol c (from the Latin word for swiftness, *celeritas*). Einstein's special theory of relativity says this speed is constant and that it represents an upper limit on the speed of everything in the Universe. However, it is worth pointing out that light can travel more slowly than this speed. Light travels at different speed in different materials – it is around a third slower in glass than in a vacuum, for example.

When do things happen?

That may sound an odd question but it is worth thinking about. Think about a thunderstorm for a moment. Since the speed of light is different from the speed of sound, you see the lightning and hear the thunder at different times. If you were deaf, you would say that seeing the lightning defined the moment the storm struck. If you were blind, you would say it was the moment you heard the thunder. If we think about events in space, such as the explosion of a supernova, we can think along similar lines. There was a famous supernova, SN1987a, observed on Earth in 1987. Yet, the star that exploded is 168,000 light years from us.

That means that the explosion actually took place 168,000 years previously and it has taken the light that long to reach us. However, someone on a planet orbiting the star would have realised within minutes that something had gone terribly wrong with their sun. This means that deciding which things happen at the same time is impossible.

Einstein's train experiment

Albert Einstein came up with a thought experiment to show why saying things happen at the same time is impossible. Imagine you are standing at the side of a train track. Meanwhile, a friend is on board a passing train, standing midway along one of the carriages. Just as your friend passes you, they send a flash of light towards the front and back of the train. From their point of view, the ends of the carriage are equal distances from them and they see the

ABOVE Like a speeding train, time is always on the move.

reflections from the front and the back walls of the carriage at the same moment. Things appear differently to you. Since the train is moving, in the time that it takes the flash of light to reach the walls, the train has moved forward slightly. This means that the reflection from the back wall (which is slightly closer to you) reaches you sooner than the reflection from the front wall (which is slightly further away). In other words, the two reflections are not simultaneous. Einstein said that this proved that saying things happen at the same time is worthless.

ABOVE Space is three dimensional, time playing the role of a fourth dimension.

The space-time continuum

Einstein's special theory of relativity was published in 1905 but it was a former teacher, Hermann Minkowski, who came up with the idea that it made more sense if the universe did not have just three dimensions (side to side, up and down and front to back) but four (those three plus one time dimension). This way, the problem of simultaneity disappeared – by referring to an event happening at a particular location in the universe at a particular time and by measuring the distance between two events as a combination of both normal distance and time.

Weird things about time

Time slows down: *If someone wearing a watch zooms away from you at a speed approaching that of light and you look at their watch through a telescope, you'll see that its hands are moving more slowly than those on your own watch. This is known as time dilation.*

Everything is relative: *If that person speeding away from you also had a telescope and looked at your watch, they would think that your watch was the one moving slowly.*

Speeds don't add up: *If two cars traveling at 60 mph (96 kph) in opposite directions on a road crash into each other, they are traveling at a relative speed of 60 + 60 = 120 mph (193 kph). This is why head-on collisions are worse than tail-end collisions. You would imagine that two spaceships which whiz away from each other at 60 per cent of the speed of light have a relative speed of 120 per cent of the speed of light. In fact, Einstein's relativity says that their relative speed is 88 per cent of the speed of light.*

Mass slows down time: *As well as speed affecting time, gravity affects time too. Einstein's general theory of relativity says massive objects warp spacetime (rather like a bowling ball would stretch a rubber sheet). This has the effect of slowing down time near heavy objects.*

GPS satellites: *The system of satellites orbiting the Earth which are used in the Global Positioning System have their clocks adjusted to take account of both speed and gravitational time dilation. If they didn't, the positioning information would be more inaccurate.*

FACT

GPS has been in use by the US military since the mid-1960s. Advances in this area have made the deployment of weapons a pinpointed science in itself, affording the opportunity to reduce unintended casualties. However, the evolution of the technology for the rest of us is more remarkable still. The next generation of navigation systems will make use of detailed imaging resources such as Google Earth, so drivers will be viewing exact and up to date images en route rather than the uninspiring and sometimes confusing vector graphics currently employed.

ABOVE A cosmic ray entering the Earth's atmosphere.

Time dilation and cosmic rays

There is a particle called the muon which particle physicists have studied extensively. It is unstable and changes (or, as the particle physicists say, decays) into an electron and two neutrinos in the blink of an eye (well faster actually, in an average of 2.2 microseconds).

Unusually, we see a lot of muons on Earth that are created by cosmic rays – high energy particles coming from outer space – which smash into particles in the atmosphere, a sort of natural particle accelerator. The odd thing is that given the average life span of a muon we should see hardly any on the surface of the Earth – they should have nearly all decayed into electrons and neutrinos. In fact, it is because the muons travel at nearly the speed of light – and suffer from time dilation – that their average lifespan is extended, allowing them to reach the Earth's surface.

Five cool things about tachyons

Tachyons are objects which may travel faster than the speed of light. They have never been observed.

Fast particles: *The name tachyon comes from tachy – the Greek for fast. They are the opposite of tardyons. Everything observed in the Universe so far is either a tardyon, or a luxon – something that always travels at the speed of light – a photon of light for example.*

Energy barrier: *Relativity says a spaceship could never accelerate to the speed of light because its mass increases exponentially the faster it goes. Reaching the speed of light would take an infinite amount of energy. Slowing a tachyon down to the speed of light would also take an infinite amount of energy.*

The variable speed of light: *Light travels at different speeds in different materials. If an electrically charged particle enters water faster than the speed of light in water, it will emit Cerenkov radiation as it slows down – the blue glow of a nuclear reactor is caused by this. Since tachyons supposedly travel faster than the speed of light in a vacuum, if one is passing by, it should emit Cerenkov radiation. It has not yet been observed by scientists.*

Seeing double: *If a tachyon were heading straight for you, you wouldn't see it since it would be traveling faster than any light that could be reflected from its surface. Bizarrely, once it had passed you, you would see double – one image that eventually reached you from where it came from and another in the direction it had gone.*

The really long ruler: *If you had a ruler that measured from here to the Sun and you moved it from your right to left hand in a couple of seconds, would that mean some method of faster-than-light communication had told the far end that it had to move at the same time? In fact, any rigid object is not rigid at all but just a collection of loosely associated atoms and molecules. The knock-on effect of the forces between adjacent atoms travels far more slowly than the speed of light.*

ABOVE The atomic clock, first built in 1949 by the US National Bureau of Standards.

How long is a second?

The second used to be defined very simply as 1/86400th of the average day (24 hours X 60 minutes X 60 seconds). However, scientists realized that the length of a day is always changing so they needed to come up with something else. Now, the second has a much weirder definition. In the words of the Bureau International des Poids et Mesures (International Office of Weights and Measures) who standardize such things, the second is now defined as "the duration of 9,192,631,770 periods of the radiation corresponding to the transition between the two hyperfine levels of the ground state of the caesium 133 atom". It turns out the electrons orbiting atoms of caesium are a far more accurate means of measuring time. The metal caesium is therefore the basis of atomic clocks.

The twin paradox

Traveling fast makes your watch go more slowly. This leads to an interesting possible scenario. One of a pair of identical twins decides to visit Proxima Centauri and zooms off in a spaceship that can travel at 99% of the speed of light. After a brief look around, they return home. Since Proxima Centauri is 4.22 light years away, eight and half years would have elapsed for the traveler. However, at 99 percent of the speed of light, time slows down by seven times compared to a stationary observer. This means that on return, the stay-at-home twin would have aged $7 \times 8.5 = 59.5$ years. The apparent problem with this (the paradox) is that if you turn the situation on its head and consider things from the perspective of the traveling twin, that the homebased twin appears to move away and then return, you would think that the stay-at-home twin should be the younger one. However, Einstein's general theory of relativity says that the trip is not symmetrical at all and the traveling twin really would be younger. Tests of this, using accurate clocks, have proven that Einstein was correct.

The grandfather paradox

Imagine you invented a time machine and used it to go back in time. Unfortunately, while driving around in the 1960s, you accidentally run down and kill your grandfather before he met your grandmother. Logically, because your grandparents would now never meet, you would not be born. Of course, this means that you would then not be able to invent a time machine and travel back to accidentally kill him. This is the so-called grandfather paradox. Some believe that because of the problems it introduces, time travel is impossible. Others argue that the act of killing your grandfather would create a parallel universe in which you never existed, although you would be able to return to your own parallel universe using your time machine.

Time waits for no-one

The concept of time is a remarkable thing, within a moment or an eternity, the wonders of the universe unfold.

A table of times	
5.3×10^{-44} seconds	Planck time, the shortest duration which scientists believe will ever be measurable.
0.00009 seconds	The amount of time it will take a proton to make a single circuit of the Large Hadron Collider.
0.0014 seconds	The period of rotation of the pulsar PSR J1748-2446ad, the fastest spinning pulsar known to date.
1.3 seconds	The time it takes light to travel from the Moon to the Earth.
43 seconds	The time it took from release to explosion of Little Boy, the atom bomb that destroyed Hiroshima.
886 seconds	The average lifetime of a neutron (when not bound inside an atom).
5,480 seconds	The time it takes the International Space Station to complete an orbit of the Earth.
35,733 seconds	The length of a day on Jupiter.
133,081,920 seconds	The length of time it takes light to reach Earth from Promixa Centauri, the nearest star after the Sun.
1.41×1017 seconds	The half-life of uranium-238, i.e. the time it takes half of a sample of the element to undergo radioactive decay.

EVERYDAY FREAKY SCIENCE

Why is the sky blue?

Sunlight is a mixture of many different colors – rainbows are caused by the splitting up of the light into its constituents by passing through raindrops. The color of the sky is determined by a process called Rayleigh scattering, the bending of different colors of light by different amounts by the particles in the atmosphere. Blue light is scattered far more than red light so you see blue light coming from all parts of the sky while the redder tones are concentrated around the disk of the sun.

Why does the sky look red at sunset?

The cause is pollution in the atmosphere. While the light from the midday sun passes straight down through the atmosphere, the sunlight at sunset comes in at an angle, meaning that its journey through the atmosphere is far longer. Particles in the atmosphere absorb light as well as scattering it, with colors other than reds and oranges being absorbed the most, explaining why you see those warm hues at day's end.

ABOVE In popular folklore a red sky at sunset was a sign of fair weather to follow.

Which way does water go down the plughole?

You may have heard that water goes down the plughole in different directions in the Northern and Southern hemispheres. Some say it is down to the Coriolis effect. Now while the Coriolis effect does exist – and is the reason why similar weather systems rotate in opposite directions in the two hemispheres – the same does not apply to water and plugholes. In fact, water will go down a plughole in a direction that is effectively random. The shape of your bath tub or sink and turbulence in the water far outweigh any influence of the Coriolis effect. Try it yourself a few times and see what happens. You can even get the water to spin the opposite way with a well-timed swish of the hand.

Can you save yourself by jumping up at the last minute in a falling elevator?

It's the nightmare scenario; the elevator stops and then suddenly the cable snaps, sending the elevator plunging to the ground. You comfort yourself that at the moment before it hits the ground, you can jump up, saving yourself from being killed. Sadly, this is not true. Both you and the elevator start to go into free fall. The floor is no longer pushing up against your feet and you start to feel weightless. As you fall, both you and the elevator increase speed. At the moment before impact, you would need to somehow reduce this speed to stop you crashing into the ground. If you have been falling for several floors already, your legs wouldn't have enough power in them to decelerate you sufficiently. Even if they did, the cabin itself would almost certainly crumple on impact, with the elevator roof squashing you flat.

How does your cell phone pinpoint where you are?

Cell phones work using cell technology. Cells are areas of cell phone coverage which are typically a few miles across and are centered on a radio mast. Your phone shows you have a signal when you are within range of this mast and radio signals pass between the phone and the mast, allowing you to make a call. So how does this help pinpoint your location? First off, the network knows which cell you are in. The strength of the signal shows how far away you are from the mast. But in built-up areas, you are often in range of several masts and the strengths of the signals from each of these can then be used to find you, using a process known as triangulation.

RIGHT Condensation forming on a can.

Why do cold cans of drink get beads of water on them?

Why do cold cans of carbonated drinks get beads of water on them when you take them out of the refrigerator? It's a process called condensation. Inside the refrigerator, the can, the liquid inside it, and the air inside the refrigerator are all at the same chilly temperature. But when you take the can out of the refrigerator, suddenly the surrounding air is at room temperature. Air contains water but you normally can't see it. The most obvious proof of this is to see rainclouds form from seemingly thin air. Getting back to our can, its cold surface chills the surrounding warmer air, causing the water in the air to turn into a liquid, which condenses on the sides of the can.

Some cool equations and what they mean

Mathematicians and scientists don't use the multiplication sign in their equations because it gets confusing. Instead, they just run all of the numbers and letters together and assume you know that means you have to multiply them together – so the right hand side of $E = mc^2$ really means m X c^2 but in shorthand.

$E=mc^2$

This is probably the most famous equation of all time, apart from $1 + 1 = 2$ perhaps. It is the equation at the heart of Albert Einstein's special theory of relativity and shows that energy (E) and mass (m) are interchangeable. The other symbol in the equation (c) is the speed of light in a vacuum. Interchangeable here means that you can convert one into the other and this is the principle involved in nuclear fusion and particle accelerators. If you measure the mass of the two protons and two neutrons which make up a helium atom, you will notice that the atom weighs slightly less than the four constituent particles on their own. The difference in mass is what is converted into energy in the process of fusion; the amount of energy involved is given by that famous formula. In a particle accelerator, particles like protons and electrons are accelerated to high speeds. When they collide with other particles, the energy due to their motion (their kinetic energy) creates a spray of new particles far more massive than the colliding objects.

$T^2=kD^3$

The German astronomer Johannes Kepler was born in 1571 and was the first to realize that the planets move in squashed ovals, or ellipses, around the Sun, rather than circles. He came up with three laws of planetary motion, the third of which is known as Kepler's third law. What the equation means is that the square of the length of a planet's year, represented by T (in Earth years), is proportional to the cube of its average distance, D, from the sun. Astronomers use something called the astronomical unit (AU) to represent the average distance from the Earth to the Sun (1 AU = 93 million miles, or just under 150 million km) so knowing that the Earth year is 1 year long, and observing how long the year of another planet is, you can work out the other planet's distance from the Sun.

So for Earth:

$$1^2 = k1^3$$

Now 1^2 and 11^3 both equal 1, so therefore $k = 1$.

For Jupiter

$$11.9^2 = kD^3$$

or

$$141.6 = 1 \times D^3$$

Taking the cube root of 141.6, gives us $D = 5.2$, so the average distance of Jupiter from the Sun is 5.2 astronomical units, or 485 million miles (780.5 million km).

Why it's so hard to push a stationary car

If a friend has an old car, you may have been asked to help give it a push start when the battery has gone flat. What you will certainly have noticed is that trying to push it when it is stationary is incredibly hard but pushing it once it has started moving is a lot easier. Anyone who doesn't realize this is often left flat on their face. In

ABOVE A car is easier to push once it is already moving forward.

fact, this is just an everyday example of Newton's second law of motion, or F=ma as it's often known. In this equation, the *F* means force, or how hard you are pushing it, *m* is the mass of the car (how heavy it is) and *a* is how fast it accelerates. What it tells us is that if you push the car with a constant amount of force, since the car's mass doesn't change, then it will be constantly accelerated, i.e. it will get faster and faster and faster, leaving you sprawling in the mud.

Newton's universal law of gravitation $F = \dfrac{GMm}{r^2}$

This equation, deduced by Sir Isaac Newton, tells us about gravity. It says that the gravitational force (F) on an object with mass *m*, as a result of another object with mass *M* located a distance (r) away from it, is equal to the two masses multiplied together, divided by the square of the distance between them, which is then all multiplied by something called *G*, the universal gravitational constant.

That's fine but what practical use does it have? Coupled with Newton's laws of motion, we can use it to measure the mass of the Earth itself (M). If we set up an experiment in which we drop a ball from a height and measure the amount of time it takes to fall a fixed distance, we can calculate how fast it accelerates because of gravity.

We can say that the force on the ball, which has a mass *m*, from gravity is equal to that from Newton's second law

$$F = \frac{GMm}{r} = ma$$

We can divide both sides by m to get:

$$a = \frac{GM}{r^2}$$

Now we can work out a from the distance the ball drops (x) and how long it took to drop (t) from another of Newton's equations of motion:

$$a = \frac{2x}{t^2} = \frac{GM}{r^2}$$

Rearranging this we get:

$$M = \frac{2xr^2}{GT^2}$$

Newton provided us with a value for G and we know what x and t are from our experiment but what about r? This is the distance between the ball and the Earth. It seems this should be zero since the ball is on the surface of the Earth but in fact gravity works as if all of the mass was concentrated right at its center, so r is therefore the radius of the Earth. There are ways to work out the radius and plugging in all the numbers, we can find that the Earth weighs 6 x 2,257 lbs (6 x 1,024 kg).

FACT
Astronauts grow several inches out in space without the force of gravity pulling them down, but back on Earth they shrink back to their original height.

Cool stuff about magnets

Why are magnets magnetic?: *All magnets ever made or seen in nature have two opposing poles – north and south. If you suspend a magnet freely, the north pole will always end up pointing north and the south pole south. Magnetism comes about because the electrons that whiz around the atoms that constitute magnets are also tiny magnets with their own poles. In non-magnetic materials, these electron magnetic fields all point in different directions, canceling out the magnetic effect.*

Single magnetic poles: *Experiments with a pair of bar magnets show that opposite poles attract each other while similar poles repel each other. If a bar magnet is painted red one end and black the other, what happens if you cut the magnet in half along the dividing line? You are not left with a single north magnetic pole and a single south magnetic pole; the two halves act just like the original magnet, each having north and south poles. In fact, single magnetic poles have never been seen, although some theoretical physicists believe they might exist, but are just incredibly rare in nature.*

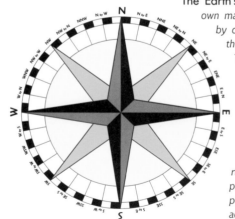

ABOVE The familiar mariner's compass was invented in Europe around 1190.

The Earth's all wrong: *The Earth has its own magnetism, believed to be caused by coordinated electrical currents in the liquid metal core of the Earth. The interaction of the solar wind – a stream of particles emerging constantly from the burning Sun – with the Earth's magnetic field causes the Northern and Southern Lights. Yet opposite magnetic poles attract.*

This means that if we stick to the rule that the end of a magnet pointing north is a north magnetic pole, then the Earth's north pole is actually a south magnetic pole. How confusing is that?

The world's strongest magnet: *Magnet strength is measured in a unit called teslas (named after the Serbian inventor Nikolai Tesla). A typical bar magnet has a strength of about 0.01 tesla while the magnetic field of the Earth is only 0.00005 tesla.*

The world's strongest magnet is at the National High Magnetic Field Laboratory in Florida. It generates a steady 45 tesla and needs to be chilled to 2 degrees above zero to operate, necessitating 4,000 gallons of cooling water every minute.

The universe's strongest magnet: *Neutron stars, which contain up to twice as much mass as the Sun but are just 10 to 20 miles (16 to 32 km) across, are the record holders when it comes to magnetic fields; strengths of up to 100 billion tesla have been spotted. Neutron stars with very high magnetic fields have now been called magnetars.*

Everyday magnets: *Magnets are used by people in countless common devices such hairdryers, telephones, vacuum cleaners, electric mowers and ipods.. Computers use magnets to save information. Huge magnets are used to separate waste. Some trains also use electromagnets, riding without actually touching the rails. In the train and in the rails are powerful magnets that repel each other. The train hovers by the magnetic power and is moved forward by it too.*

A timeline of computing and technology

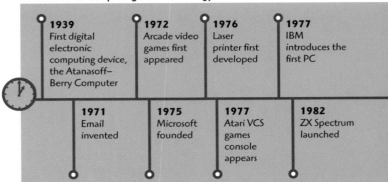

1939 First digital electronic computing device, the Atanasoff–Berry Computer	1972 Arcade video games first appeared	1976 Laser printer first developed	1977 IBM introduces the first PC
1971 Email invented	1975 Microsoft founded	1977 Atari VCS games console appears	1982 ZX Spectrum launched

Top five most powerful computers

Every six months, an organization called Top500 produces a list of the top 500 supercomputers around the world. The power of a supercomputer is measured in FLOPS – floating point operations per second – basically a measure of how many calculations in can carry out in a second.

1. **Lawrence Livermore National Laboratory:** *An IBM supercomputer called BlueGene/P, used to model the US nuclear weapons stockpile. It has a sustained power of 478.2 trillion FLOPS (teraFLOPS).*

2. **Forschungszentrum Juelich (FZJ):** *Also an IBM BlueGene computer but a P variety, capable of a sustained power of 167.3 teraFLOPS. It is located at a former nuclear research center in Juelich, Germany.*

3. **NMCAC:** *Has a speed of 126.9 teraFLOPS and is based at the New Mexico Computing Applications Center (NMCAC) in Rio Rancho. It was built by SGI and is used for general supercomputing research needs.*

4. **Computational Research Laboratories:** *Based in Pune, India and is operated by the huge conglomerate Tata Sons. The HP machine achieved a top speed of 117.9 teraFLOPS.*

5. **Swedish Government Agency:** *This achieves a top speed of 102.8 teraFLOPS. It is used for unspecified military purposes.*

1983 Microsoft releases Word and Windows	**1984** Apple launches the Macintosh	**1990** First digital mobile phone call	**1994** First Sony Playstation launched	**2003** MySpace.com founded
1983 First commercial mobile phone, the Motorola DynaTAC 8000X	**1989** The World Wide Web is invented	**1993** Netscape web browser launched	**2001** Apple launches the iPod	**2007** Apple launches the iPhone

Top five feats of miniaturization

1. **The cell phone:** *The first cell phones were enormous bricks that needed a battery the size of the full set of Harry Potter books to get enough power to speak for just a few minutes. Today's tiny phones contain cameras, can run email and sat nav applications, and can remain charged for days.*

2. **Computer chips:** *When chip maker Intel introduced the 8086 processor more than 25 years ago, it contained just 29,000 transistors. Intel's latest Pentium chips cram hundreds of millions into a similar amount of space.*

3. **Laptops:** *Many computer manufacturers claim to have come up with the world's smallest laptop but some are little more than glorified phones. Two that really stand out are the Asus Eee, with its 7-inch (18-cm) screen and weight of less than 2 lbs (1 kg), and the Macbook Air, so thin that Apple boss Steve Jobs pulled it out of an envelope when he first announced it.*

4. **World's smallest guitar:** *In 1997, nanotechnology experts at Cornell University carved a guitar that was just 10 micrometers long out of a crystal of silicon. Each of its six strings was about 50 nanometers wide, about the width of 100 atoms.*

5. **World's smallest robot:** *In 2005, researchers at Dartmouth built a robot that was as wide as a human hair and 250 micrometers long. It couldn't do much other than move along in tiny steps but it was controllable.*

ABOVE The idea for an integrated circuit was recorded as early as 1958 by Jack Kilby, for which he won the ultimate recognition of a Nobel Prize in 2000.

What is nanotechnology?

Nanotechnology is technology on the scale of nanometers, one 50,000th of the diameter of a human hair. The miniaturization techniques that have given us super powerful computers are now reaching this scale. However, this is the realm of quantum physics and traditional methods of making things ever smaller just won't work.

ABOVE Nanobot in action.

Nanobots

Nanotechnologists are getting very excited about nanobots (robots on the nanotechnology scale). Although largely science fiction at the moment, scientists think that when production problems are overcome, nanobots will be capable of swarming around the body fixing cancer cells or performing internal surgery with incredible precision. Some even think that armies of nanobots will be used by the military to wage an unseen war on their enemies.

Gray goo

The term gray goo was first used in a book by nanotechnologist Eric Drexler in 1986. It refers to a Doomsday scenario for the world and perhaps the universe involving self-replicating nanobots. If this self-replication gets out of control, there are people who worry that they will consume every available resource on the planet, reducing it to a swarm of nanobots, with an overall gray and gooey consistency. Yuck.

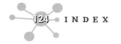

INDEX

6

X

Y

PICTURE CREDITS

The publishers would like to thank the following for permission to reproduce pictures.

Corbis: p. 40; Dreamstime: pp.12, 25, 51, 52, 53, 68, 73, 74, 75, 93, 96, 103, 107, 112, 114; Getty: p. 95; iStockphoto: pp. 13, 23, 30, 48, 69, 76, 87, 111, 113, 116,117, 119, 123; Korea Times: p. 99; NASA: pp. 30, 31, 34, 35, 36, 37, 43, 44, 45, 49, 50, 54, 55, 56, 58, 59, 60, 63, 71, 117; Science Photo Library: pp. 26, 29, 66, 67, 77, 82, 90, 100, 101, 104, 105, 106, 108, 109; Wikipedia: pp. 9, 14, 17, 19, 24, 67, 68, 83, 89, 91, 120, 122

Illustrated by:

David Eaton: p. 119; Richard Burgess: pp. 18, 32

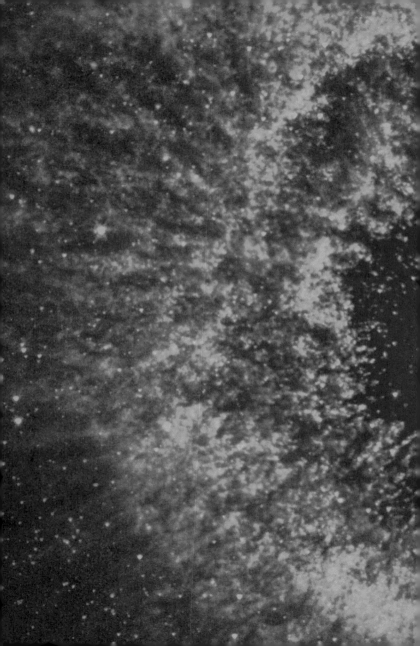